KID TECH

Hands-On Problem Solving

with Design Technology

for Grades 5–8

LUCY MILLER

Dale Seymour Publications®
Menlo Park, California

About the Author: Lucy Miller is from Nova Scotia and works with the Carleton Roman Catholic School Board in Ottawa, Canada. She has been in education for twenty-three years as a teacher, curriculum consultant, principal, and part-time professor at the University of Ottawa in the Faculty of Education. Lucy has published a series of thematic math problems and gives workshops in design technology, mathematics, reading, and writing. She has won the Quality in Education Award from the Nova Scotia Teachers' Union, the Director of Education's Award, and The Prime Minister's Award for Teaching Excellence in Science, Technology, and Mathematics.

Managing Editor: Catherine Anderson
Production/Manufacturing Director: Janet Yearian
Design Manager: Jeff Kelly
Project Editor: Joan Gideon
Senior Production Coordinator: Alan Noyes
Art: Rachel Gage
Interior and Cover Design: Polly Christensen, Christensen & Son Design
Composition: Brandon and Penny Carson

Art on pages 5, 28, 30, 34, 36, and 38 was first published in *TTS Design and Make Activity Folios* and is reproduced here courtesy of Steve Rogers, Technology Teaching Systems.

Art on pages 4–6, 30, 35, 53, and 64 was first published in *Techniques for Technology* and is reproduced here courtesy of Steve Rogers, Technology Teaching Systems, Inc.

Many of the designations used by manufacturers and sellers to distinguish their products are claimed as trademarks. When Addison Wesley Longman was aware of a trademark claim for such designations as appear in this book, the designations have been printed with initial capital letters.

Published by Dale Seymour Publications®, an imprint of Addison Wesley Longman, Inc.

Dale Seymour Publications
2725 Sand Hill Road
Menlo Park, CA 94025
Customer Service: 800-872-1100.

Order Number DS29705

ISBN 1-57232-541-0

1 2 3 4 5 6 7 8 9 10-ML-02 01 00 99 98 97

This Book Is Printed
On Recycled Paper

CONTENTS

INTRODUCTION

TECHNOLOGY IS HUMAN INNOVATION IN ACTION; it involves the generation of knowledge and processes to develop systems that solve problems and extend human capabilities." (Technology for All Americans Project, *Technology for all Americans: A Rationale and Structure for the Study of Technology*. Reston, Virginia: International Technology Education Association, 1996). In the classroom, this means the use of knowledge, tools, and resources to produce the goods and services that solve a problem or fill a need.

We live in a technological world, and technology is our means of sustaining and improving it. If today's students are to become successful, contributing members of society, they must be knowledgeable about technology and able to apply it in diverse situations. To be technologically literate, students must be confident using technology to solve real-life problems, to communicate, and to work with others to provide goods and services for themselves and others.

Design technology integrates problem solving with an awareness of the effects of technology on society and the environment. It puts problem solving in a realistic context, and it addresses situations that are meaningful to students. Its goals are to develop confidence in problem solving and competence in using technology wisely.

Design technology is not a subject to be taught in isolation. By its nature, it lends itself to integration across the curriculum. It should be viewed as a vehicle for addressing the goals of existing programs in a more authentic and efficient manner.

GETTING STARTED

GETTING STARTED IS AN INTIMIDATING THOUGHT for some teachers, but it need not be. This resource offers a practical step-by-step approach for using design technology to accomplish existing educational goals. Design technology can and should be an integral part of the existing curriculum. As such, it offers many ways to enhance all subject areas throughout the school year.

- The problem-solving method used in the design process develops critical thinking and creative problem-solving skills.

- The design team approach shows students the value of working cooperatively.

- Students integrate knowledge, skills, and values from a variety of subject areas.

- Students learn to read and write technical material and conduct research in a context that is meaningful to them as citizens of their technological world.

- Students develop communication skills through dialogue and product presentations.

- Students improve their vocabulary by sharing knowledge and skills with others and through exposure to new vocabulary in the design briefs.

- Design brief recording sheets model and develop the organizational skills required to approach a problem.

- Visual, tactile, and auditory learning styles are addressed.

- Individual, small group, and whole group learning opportunities are provided.

- Problem solving is open-ended, allowing for a variety of solutions.

- Students are consistently involved in authentic roles and authentic learning.

- There is an ongoing focus on the similarities between various cultures since technology is often used in the same or similar ways by many cultures.

To integrate design technology into the curriculum, follow these steps.

1. Set up the classroom—Suggestions for how to do this are provided on page 3.

2. Organize design teams and teach the design process—Strategies for setting up design teams as well as a detailed description of the design process and the roles and responsibilities of design team members are provided in the Design Process section on page 8.

3. Introduce the tools—A description of the necessary tools and instructions on how to use them safely are provided on page 4.

4. Choose the design briefs—This choice should be determined by the units or themes being covering in class. Design briefs should be planned to ensure the integration of cross-curricular learning outcomes or performance standards. Increased integration provides greater opportunity for students to see the

links between the knowledge, skills, and values they have learned. Learning becomes meaningful when it relates to what students already know and to the real world. Design technology is an efficient vehicle for doing this. Every design brief facilitates several technology learning outcomes and addresses performance standards in several disciplines. Science, mathematics, language, social studies, and the arts are all addressed either directly through the briefs or through the extensions suggested. The briefs can be used to introduce a topic or theme; to follow up or introduce a piece of literature; to assess a student's understanding of a concept in science, math, social studies, or the arts; to replace traditional science fair projects; or to replace a complete unit on a concept or theme.

5. Teach basic techniques—Before assigning a design brief, teach the basic techniques that will be needed for that brief. These will be listed in the teacher notes for the brief and are provided in the Basic Techniques section on page 28.

6. Assess student performance—Assessment is not limited to assessment of the finished product. Assessment of the problem-solving process and of group and individual performance as well as self-evaluation by students should be ongoing. Guidelines for assessment are provided starting on page 14.

SETTING UP THE CLASSROOM

Collecting and Storing Material

Many of the materials needed to complete the design briefs can be found in school art, math, or science supplies. Some things will have to be purchased. Supplies can be built up gradually. It is important to remember that design technology encourages using available materials to solve real-life problems. Be creative in substituting available

materials for those suggested in the design briefs. Many of the materials referred to in the design briefs can be gathered by students at home or donated by local businesses. A list of materials to gather has been provided, as well as a sample letter that could be used to encourage parents' participation.

Organization and storage of materials can be done at the school or classroom level, depending on availability of space. Sometimes a movable storage unit containing tools and materials can be located centrally and rolled into a classroom as needed. It is important that the materials are organized, clearly labeled, and accessible to students when they are working on their design briefs. Choosing the materials is an essential skill to learn in design technology. Students must also learn to value the materials, to save scraps, and to keep the materials organized at all times. Involving students in collecting many of their own materials is an effective means of developing this sense of responsibility.

A Technology Center

If there is room for a technology center in the classroom, it should include as many of the items as possible from the list that follows on the next page.

- tools

- materials gathered by students from home and local businesses

- construction, movement, and power kits

- books—fiction that lends itself to technological discussion, such as most fairy tales and science fiction; biographies of inventors; non-fiction books on science, technology, and theme-related topics; magazines, tool catalogues, and catalogues of electrical, plumbing, and computer supplies

- broken toys, clocks, small appliances, and other items to take apart

- safety posters and posters advertising student inventions

- a list of class-generated rules

- a list of the roles and responsibilities of design team members

- samples of student work at various stages of completion—these should include finished products as well as design brief recording sheets, drawings and sketches, and quality criteria.

- computer, camcorder, and tape recorder

Tools

Students need to be able to use a variety of tools as they carry out solutions to the design briefs. Students should learn how to use the following tools.

- **Bench Hook or Miter Box**—These are placed over a desk or table to provide a solid surface for cutting and allow you to hold the wood strips in place for cutting. The miter box looks just like a bench hook but has grooves that are used to guide the saw for straight or pre-measured angled cuts. A miter box is a better investment because it can also serve as a bench hook.

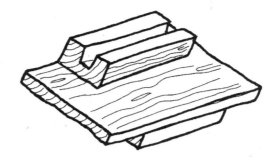

- **Hammer**—This should be a small ball peen hammer, rather than the large carpenter's hammer. Students should be taught how to hold the hammer handle as far away from the hammer head as possible to ensure greater leverage.

- **Saw**—Small hacksaws are easy to use and safe for student use. Blades must be sharp and easy for students to replace. To start a cut, the saw should be pulled backwards a few times to make a cutting groove. Press gently and use the whole blade.

- **Drill**—A hand drill with no exposed mechanisms is safe and works well. Students should be taught to press lightly when turning the handle and to use a scrap of wood under the material being drilled. Have students work in pairs so a partner can hold the material steady. Instruct students to turn the handle the opposite way while pulling lightly to remove the drill.

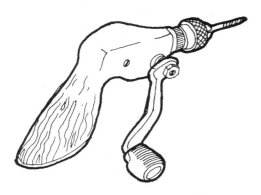

- **Jointer**—A jointer is usually made of strong plastic and holds two pieces of wood at right angles for gluing. Use of a jointer takes the frustration out of making square corners.

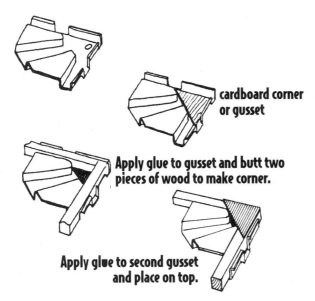

cardboard corner or gusset

Apply glue to gusset and butt two pieces of wood to make corner.

Apply glue to second gusset and place on top.

- **Glue Gun**—Low temperature glue guns are safe and useful for gluing plastic to wood or metal, since the glue dries quickly. Regular glue guns can be used by the teacher or by older students with supervision.

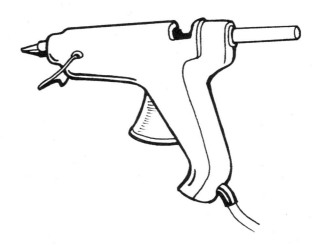

- **Cutting Mat, Utility Knife, and Metal Safety Ruler**—There should be a cutting mat available whenever students work with a utility knife. A cutting mat can be made of wood, heavy cardboard, or plastic. Kitchen cutting boards are ideal. The utility knife should always be used with a raised metal ruler to guide the cut and protect hands. Instruct students to keep hands inside the indentation guide and to retract the blade when it is not in use.

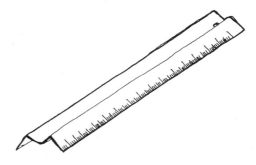

- **Safety Glasses**—Whenever necessary, students should wear safety glasses. They are inexpensive, and each student should bring a pair from home or be provided with a pair at school.

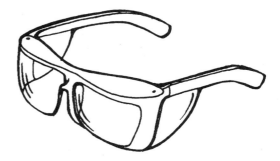

- **Hole Punch**—A single-hole punch will often be needed to make holes in cardboard, ice cream sticks, paper, or cloth. Technology suppliers provide ones that are sturdier than the ones previously used in classrooms.

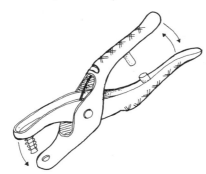

- **Wire Strippers**—When doing electrical work, wire strippers are needed to strip the covering from wires. They usually contain a craft blade, which can be used to cut wire. Students should be taught never to put their fingers near this blade!

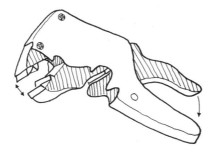

- **Compass Cutter**—A compass cutter can be used to cut circles of different diameters in paper, cardboard, and light plastic. It has sharp blades like a utility knife and should always be used on a cutting mat. Only adults should use the compass cutter.

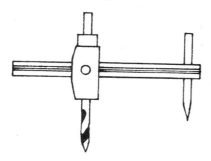

- **Screwdrivers**—A variety of screwdrivers will be needed for taking things apart and putting things together. Students should be taught that certain screwdrivers are used with certain types of screws. They should also learn that using a small nail or drilling to make a small entry hole will prevent wood from splitting and allow the screw to go in more easily.

- **Snips**—Snips work like scissors to cut plastic, tin, tubing, or thin wooden sticks.

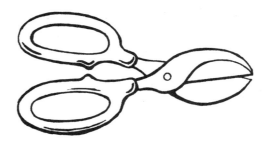

- **Table Vise**—A vise can be attached to a desk or table to hold materials as they are being worked on, or after they have been glued.

- **Adhesives**—A wide variety of adhesives should be available. There will be opportunities to explore their properties in the design briefs. Some of the most commonly used are white paper glue, carpenter's glue, adhesive tapes, and glue sticks.

Materials List

paper shopping bags
lollipop sticks
heavy boxes (cut into pieces)
meat and fruit trays
fabric remnants
plastic containers and bottles
straight pins and safety pins
wheels
wooden dowel
small appliances (broken)
broken clocks or windup toys
metal and wooden skewers
ice cream sticks
eye hooks
rubber bands
stir sticks
stopwatch
pipe cleaners
nails
aluminum pie plates
wooden strips

cardboard tubes
sponges
fishing line
clothespins
string, thread, and yarn
copper or aluminum foil
film containers
small pieces of wood
old building sets
Styrofoam
twist ties
beads and buttons
plant and garden sticks
balloons
plastic tubing and straws
magnets and springs
coat hangers and wire
paper fasteners
candles
poster board
paper plates and cups

gift and food boxes
rubber washers
panty hose cards
greeting cards
milk cartons
jars, cans, and lids
nuts and bolts
old toys and games
sandpaper
mirrors
eggbeaters
knitting needles
bobbins and spools
syringes without needles
corks
spring scale
clay and sand
paper clips
thread
combs
spatulas

Sample Letter to Parents

Dear Parents,

Our students will soon begin investigating the impact of technology on our world. Our study of technology will be integrated with our study of math, science, language, the arts, and social studies. It will involve problem-solving with an awareness of the effects of technology on society and the environment. The problems presented will address situations that are meaningful to students.

As our partners in education, we hope that you will ask your child about what we are doing. We are also including a list of some of the materials we will be using. If you can contribute any of these items we will put them to good use. Thank you for your continued cooperation and support.

Sincerely,

THE DESIGN PROCESS

The design process uses a problem-solving method similar to the ones students have already used in mathematics and science. There are four steps.

1. **Clarify the Problem**   Students discuss the problem after it has been presented. They analyze the scenario that has been described and clarify what is being asked. It is during this step that students redefine the situation, and what they are being asked to do, in their own words.

2. **Generate Ideas for Possible Solutions to the Problem**   At this stage, students share previous knowledge and ideas on the topic. They research relevant information and any new concepts, and they make initial sketches of possible solutions. They should be encouraged to suggest many ways of arriving at the solution and to build on each other's ideas. They should also focus on using their research to revise or refine their ideas and sketches. The design process is cyclical, often requiring students to go back and rethink their approach.

3. **Implement a Solution to the Problem**   Students must now attempt the solution they have chosen. This usually involves making or building a product. Students must use the materials provided or choose from other available materials. During this step, students are faced with additional problems and difficulties as they attempt to build their product. This sometimes results in a change of plans as they realize their initial idea is not going to work. Student journal writing can document the process of adjusting the solution as problems arise.

4. **Test the Solution and Discuss the Results**   During this step, both the product and the process used are assessed by the group and the teacher. Results of the test should be communicated to other groups. This may be done through a product demonstration or trade show illustrated with students' technical drawing and writing. Evaluations should include group discussion as well as feedback from peers and the teacher.

Setting Up Design Teams

Setting up design teams in the classroom requires the same planning and considerations necessary for other small-group activities. Four students is an ideal team size, but some situations dictate groups of two or three. Groups can be of similar or mixed ability and can be chosen by the teacher or the students. Design teams can remain the same or change for each project, although students should change roles from one project to the next. The most important factor in establishing effective design teams is ensuring that student roles and responsibilities in the group are clearly outlined. Students should understand that they will be assessed both as individuals and as members of a team. The following roles and responsibilities show one method of organizing design teams. Teachers are encouraged to consider their unique situations and revise these roles and responsibilities accordingly. Students should be given a written description of the roles and responsibilities, or a list should be posted in the classroom.

Design Team Members: Roles and Responsibilities

Project Manager

The project manager oversees the whole process. The product manager will

- read the design brief aloud to other members of the team

- lead the team in clarifying the problem

- periodically bring the team together to review the given task, the agreed-upon plan, the progress they have made, and work remaining to be done

- encourage team members to report on their individual progress and describe their needs at regular team meetings

- remind all team members of their roles and responsibilities

- include all team members in discussions and encourage them to speak and explain their ideas clearly

- praise individual team members' contributions

- represent the team in interactions with the teacher

Materials Superintendent

The materials superintendent ensures that the team has the materials necessary to complete the project. The materials superintendent will

- with the help of the team, make a list of all the materials, including books and reference materials, that will be needed to complete the project

- organize collection of materials by team members at school and at home

- supervise wise use of materials and the return of unused materials

- supervise cleanup of the site and safe storage of materials and unfinished projects

Data Engineer

The data engineer supervises and records the results of all plans, trials, and tests. The data engineer will

- take notes during the early stages of the design process while team members attempt to clarify the problem and generate ideas for possible solutions

- encourage team members to generate ideas

- keep a record of the team's plan, sketches, test or trial results, and discussions

- give research assignments to all team members and coordinate the sharing of any information gathered

- keep a copy of all research

- supervise the writing and editing of all reports and illustrations

- assign responsibilities for the final presentation of the completed work among team members and supervise the presentation

Construction Engineer

The construction engineer oversees the building of the project or model. The construction engineer will

- allocate the tasks necessary to build the project or model

- ensure that the work stays on schedule by encouraging team members to use their time wisely

- ensure that safety procedures are followed

- monitor noise levels on the construction site

- encourage all team members to participate and recognize their contributions

- approve the final product before it is handed in or presented

Teaching the Design Process

Once design teams have been established, each team should begin a design technology portfolio. A three-ring notebook can be used to keep a record of work in progress. The notebook can also include handouts students will refer to on a regular basis and forms they must complete. These include

- a description of the roles and responsibilities of design team members

- the design brief recording sheet

- the product and process assessment rubrics

- any other assessment tools being used

For individual assessment, it is useful to have students keep personal journals. In this journal, they record what they do each session and keep their own sketches, research, rough drafts of the recording sheet, and rough drafts of their final product presentation.

Introduce the design process to the class as a whole. Work through the practice brief (page 27) with the entire class, using the teacher notes (page 26) as a guide. Discuss what happens at each stage of the process and what each team member should be doing as detailed below. An additional design brief may be given to the class as a whole group or used as an opportunity for teams to practice the design process, coming back to the whole group after each stage of the process to compare notes.

The Design Process: Who Does What?

Clarify the Problem

1. The project manager reads the design brief aloud to the other team members.

2. The project manager reads the design brief aloud a second time. The data engineer takes notes on important facts, questions, or points for clarification, using the design brief recording sheet as a guide.

3. The project manager leads a discussion of the problem. All team members participate in clarifying the scenario, the problem to be solved, and the product needed to solve the problem. The data engineer records important discussion points and any decisions that have been made.

4. The project manager represents the team in interactions with the teacher.

Generate Ideas for Possible Solutions to the Problem

1. The data engineer encourages team members to generate ideas for solutions by asking questions such as, What do you think we

have to do? Can we draw or sketch what we think the product should look like? What do we need to know? Where can we find this information? What kinds of materials might we need?

2. The data engineer records answers and comments. The materials superintendent makes a list of needed materials as described by the team.

3. The data engineer assigns necessary research to individual team members and coordinates sharing of the information they gathered. The data engineer encourages the team to think about whether the new information affects the initial discussions about the product, asking questions such as, Will the product we originally suggested still work? Do we want to revise our sketches or drawings? What is our final decision? What are we going to make?

4. The materials superintendent organizes the collection of materials, asking, Can these materials be found at school? Do we have to bring anything from home?

Implement a Solution to the Problem

1. The materials superintendent makes sure the team members have everything they need.

2. The construction engineer allocates the tasks needed to build the product, ensures the work stays on schedule by encouraging team members to use their time wisely, and encourages everyone to participate, praising their contributions. The construction engineer also monitors noise on the construction site and monitors safety procedures.

3. The project manager calls team meetings. This can be done at the beginning of each session or as needed. At these meetings, the project manager encourages individual team members to report on their progress and describe their needs, questions all team

members and encourages them to clearly explain their ideas, praises team members' contributions, and reminds team members of their roles and responsibilities. The data engineer reports and records all new information.

4. The materials superintendent supervises the wise use of all materials and the return of any unused materials, cleanup of the site, and safe storage of unfinished projects and materials in use.

Test the Product and Discuss the Results

1. When the product is completed, the project manager calls a team meeting. The construction engineer organizes a product test to be done at the meeting. The data engineer keeps a record of the test results and the discussion. The project manager leads the team through a discussion of the product and process, asking questions such as these. Does the product satisfy the requirements of the design brief? Are there any modifications to be made? Is there anything that can be done to improve the product? Were there any stages in the design process where our team could have performed better? Would we change anything if we were doing it again?

2. The data engineer supervises the writing and editing of final reports and illustrations, assigns responsibilities for final presentation of the completed work, and supervises the final presentation.

Design Brief Recording Sheet

Team Members _____

Clarifying the Problem

What do we have to do? | Standards we must meet

Generating Ideas for Possible Solutions to the Problem

Sketches | Things to find out or materials we need

Solving the Problem

Tasks | Assigned to | Comments by team members

Solving the Problem

Tasks	Assigned to	Comments by team members

Testing the Product and Discussing the Results

Trials	Discussion

What modifications were needed?

What improvements could be made:

How could our team have performed better?

What changes would we make for next time?

PLANNING AND ASSESSMENT STRATEGIES

A clear plan that indicates the curriculum areas you want to cover and the skills, knowledge, and values you want students to learn will create a focus for assessment. There are many opportunities for cross-curricular integration in mathematics, science, social studies, language, and art. Some of these are mentioned in the design briefs or extensions. You will think of others as you plan. In addition, you will want to assess specific technological knowledge, skills, and values, critical-thinking skills, and creative thinking as suggested below.

Technological Knowledge

Structures

- examining the characteristics of structures in the real world
- designing and building structures

Materials

- exploring the properties of different materials
- choosing appropriate materials

Fabrication

- making structures or products from given or selected materials

Mechanisms

- recognizing what causes movement
- using appropriate materials and technical knowledge to cause movement

Energy and Power

- recognizing that a source of energy is needed to enable a mechanism to work
- understanding and using different power sources

Control

- recognizing the need for regulating the operation of a mechanism
- understanding and using different means to control operations

Systems

- recognizing combinations of interrelated mechanisms that function as a system to perform a task (for example, a car)
- exploring, designing, and constructing systems to solve problems

Function

- understanding the function of products, components, and systems
- designing products or systems that function according to given specifications

Aesthetics

- recognizing and appreciating the design appeal of structures or products
- striving to design or construct structures or products that are aesthetically appealing

Ergonomics

- examining the efficiency and safety of systems or products in relation to the people who will use them
- creating systems or products that perform safely and efficiently

Awareness of Careers

- recognizing possible career choices within a technological society

Technological Skills

Design—using the design process to solve a problem

Construction—making a product or structure

Assembly—putting things together

Dexterity—manipulating objects, tools, and materials

Use of Tools—using tools correctly

Awareness of Safety—observing safety precautions when using tools and materials

Technological Values

The values developed through design technology are important in all areas of life. They are essential to participation as a contributing member of any society.

Respect

- for people we work with: recognizing their strengths, their limitations, and their contributions

- for the environment: recognizing the effects of our actions and the products we use

- for materials: avoiding waste and recognizing its impact on the needs and rights of others

Resourcefulness

- using what we have to solve a problem

Responsibility

- accomplishing an assigned task independently

- doing our share

- being accountable for our actions and their effects on others and the environment

Cooperation

- recognizing the value of working as part of a team

- working effectively as a team member

Perseverance

- continuing with a task despite difficulties

Risk Taking

- undertaking a task with confidence, knowing that if one approach is not successful, another can be tried

Critical Thinking Skills

Critical thinking is the ability to

- demonstrate comprehension of what is read or spoken

- recognize the ambiguity, contradictions, credibility, and potential in what is read or spoken

- make judgments and formulate hypothesis or conclusions based on information given or acquired through research

These critical thinking skills are addressed in design technology.

Comprehension

- of relationships, facts, and principles

- to interpret, predict, estimate, and restate

Application

- of rules, ideas, and known information

- to construct, demonstrate, and contrast

Analysis

- of organizations, structures, principles, and content

- to determine relevancy and hypothesize

Synthesis

- to reorder, combine, design, create, formulate, and modify

Evaluation

- of values, criteria, ideas, and personal performance

Creative Thinking Skills

Creative thinking is the ability to

- be open-minded when answering questions or solving problems
- take risks when answering questions or solving problems

These creative thinking skills are addressed in design technology.

Fluency

- in the assimilation and use of new meanings

Flexibility

- in being open to new ideas
- in analyzing, simplifying, rearranging, and applying ideas

Originality

- in creating new ways and suggesting changes
- in improving, proposing, designing, and creating products

Elaboration

- by discovering and using patterns to solve problems
- by editing and revising work in progress
- by completing the task

Assessment

A wide variety of techniques are the key to effective assessment and evaluation of student-centered activities. They should include

- Observation of individuals and groups of students at work
- Interaction with individuals and groups of students. Open-ended questions should be used to encourage students to explain or give an opinion that demonstrates how they are thinking or approaching a problem.
- Listening to students as they interact in their groups. Can they give logical reasons for doing things a certain way or using a particular approach? Do their comments demonstrate understanding of the skills, concepts, and values being assessed?
- Looking at the outcome of the student's or team's efforts. This should include a close examination of written work, the product, drawings, plans, sketches, and failed attempts. It should go beyond initial impressions and aesthetic presentation. Assessment could also include a media or live presentation of the finished product.

A wide variety of cross-curricular knowledge, skills, and values will be addressed in each design brief or through the extensions suggested. Do not try to formally assess every one of them. Certain ones have been targeted in each brief. You may wish to use these or identify others based on those outlined in your system.

Create an outline of what you will assess before you begin. You may wish to do this in an integrated plan similar to the one in the sample assessment pages below. Students should be involved in their assessment. This can be done by establishing quality criteria for areas you plan to assess. Students should be involved in deciding on the criteria. Examples of criteria have been included in the sample assessment. These should be used as guidelines as you brainstorm a class-generated list. The criteria should be consistent with the performance standards specific to your school. Example product and process assessment rubrics show how the students' level of competency in the technological skills and concepts may be more easily identified.

(See pages 20, 21.) Students should be encouraged to assess their own progress continuously, using the criteria that have been established. The focus should always be on asking, How could I improve this? Is there another way to do this?

It is important to keep in mind that learning is a developmental process and that students will be at different stages of development. Not every student will meet the learning outcome with the same level of competency.

Sample Planning and Assessment

Project Treasure Box

Design and construct a box with an attached lid. The box will be used by a five-year-old to store tiny treasures. It should be attractively decorated, and the completed dimensions of any side of the box should not be more than 5 inches.

♦ **Materials**   wooden strips, poster board, triangle corners, glue, art supplies, ruler, saw, miter box

Sample Finished Product

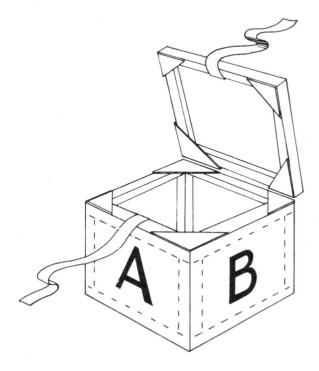

Areas to Assess

♦ **Mathematics**   linear measurement

♦ **Language**   reading and writing technical material

♦ **Art**   creating aesthetically pleasing objects, sketching

♦ **Technical Knowledge**

  • **Structures**—designing and building

  • **Materials**—choosing appropriate materials

  • **Fabrication**—making structure or products from given materials, designing products or systems that function according to specifications

  • **Aesthetics**—recognizing and appreciating the physical characteristics that make structures or products appealing; designing or constructing structures or products that are aesthetically appealing

  • **Ergonomics**—designing systems or products that are efficient and safe for the people who will use them

♦ **Technological Skills**

  • **Design**—using the design process to solve a problem

  • **Construction**—making a product or structure

  • **Dexterity**—manipulating objects, tools, and materials

  • **Use of Tools**—using tools correctly

  • **Awareness of Safety**—observing safety precautions when using tools and materials

♦ **Values**

  • **Resourcefulness**—using what we have to solve a problem

  • **Responsibility**—accomplishing an assigned task independently; doing our share; being accountable for our actions and their effects on others and the environment

Sample Assessment of Treasure Box Project

What Is Assessed?	What Standard Is Used?
1. Design Brief Recording Sheet	Quality criteria for recording sheet
2. Values	Quality criteria for group work, journal writing, and student self-evaluation checklist
3. Art	Quality criteria for product, sketches, and technical drawings
4. Language	Comprehension of design brief; quality criteria for journals
5. Math	Quality criteria for measurement
6. Product	Quality criteria for product; product rubric
7. Process	Process rubric
8. Critical thinking skills	Observation of product and process
9. Creative thinking skills	Observation of product and process

- **Cooperation**—recognizing the value of working as part of a team; working effectively as a team member
- **Perseverance**—continuing with a task despite difficulties
- **Risk Taking**—trying something with confidence, knowing that if it does not work, another approach can be taken

◆ **Critical Thinking Skills**   comprehension, application, analysis, synthesis, evaluation

◆ **Creative Thinking Skills**   fluency, flexibility, originality, elaboration

Quality Criteria

Work with students to establish the quality criteria for the class. Some of the quality criteria will apply to all design briefs. These might include the quality criteria for journal writing, the design brief recording sheet, group work, and sketches. Once you establish these quality criteria, you can post them or have students include them in their design technology portfolios. Other criteria should be established for specific projects and should be based on your school's standards of performance. Here are some suggestions for quality criteria.

Quality Criteria for Journal

- The journal is neatly written.
- The journal describes what was accomplished each time.
- The journal tries to explain why some things did not work.
- The journal includes ideas, suggestions, or questions raised during the process.
- The journal includes comments on how the team worked.

Quality Criteria for Design Brief Recording Sheet

- All sections are completed.
- Work is neatly done.
- Spelling is accurate.
- Sketches show all views of the product.
- Contributions from all team members are included.

Quality Criteria for Group Work

- Members performed their roles.
- Members listened to each other.
- Members encouraged each other.
- Time was used wisely.
- Members worked quietly.
- Members did their best.
- Members persevered when faced with problems.

Quality Criteria for the Product

- A box with an attached lid has been constructed.
- The box would be attractive to a five-year-old.
- The art work is attractive.
- All sides of the box measure 5 inches or less.
- The box is sturdy.
- The workmanship is neat.
- The box rests on a flat surface without wobbling.

Quality Criteria for Sketches

- The sketch shows all views of the product.
- The sketch is labeled to show measurements.
- The sketch is labeled to show materials.
- The sketch is neatly and accurately drawn, and a ruler is used when necessary.
- The sketch is done in pencil.

Quality Criteria for Measurement

- The student measures with precision, using the appropriate tools and units of measure.
- The student demonstrates an understanding of the differences between estimation and precise measurement.

What follows are suggested rubrics for product and process assessments.

Product Assessment

Superior Performance

The design team completed a product that was beyond the requirements of the design brief. The product satisfied all specifications and included additional features as well. Every aspect of the product was of very high quality.

Proficient Performance

The design team completed a product that satisfied all the requirements of the design brief. The product adhered to the specifications. Every aspect of the product was of high quality.

Adequate Performance

The design team completed a product that satisfied most of the requirements of the design brief. The product adhered to most of the specifications. The product was of acceptable quality.

Limited Performance

The design team partially completed a product that satisfied some of the requirements of the design brief. The product adhered to some of the specifications. The product was of inferior quality.

Process Assessment

Stages	Limited *The Student:*	Adequate *The Student:*	Proficient *The Student:*	Superior *The Student:*
Clarify the problem	Discusses the problem scenario Describes the product suggested	Discusses the problem scenario Asks questions that clarify the scenario Describes the product Asks questions that clarify the product	Discusses the problem scenario Asks questions that clarify the scenario Suggests a product	Discusses and analyzes the scenario Redefines task and product in own words Suggests some products
Generate Ideas for Possible Solution	Creates first sketch or model of product Revises sketch when relevant information and scientific concepts are reviewed by teacher Creates final sketch or model	Creates first sketch or model of product Researches relevant information and scientific concepts, using provided resources Revises first sketches, based on research	Creates first sketch or model of product Uses resources in the school to research information and scientific concepts Revises first sketches or model, based on research Discusses why changes are needed	Creates flexible first sketch Uses school resources and seeks others for research Adapts, revises sketches or model, based on research Reviews possible solutions from revised sketches or models Demonstrates understanding of concepts Chooses a practical solution
Implement Solution	Uses materials provided to make the product Makes the product	Chooses from materials provided to make the product Makes the product	Plans how to make the product Decides on materials Makes the product	Plans how to make the product Anticipates difficulties Plans to overcome difficulties Makes the product
Test and Discuss	Tests the product Communicates the results	Tests the product Discusses the product Communicates the results	Tests the product Discusses the process and product Clearly explains the results	Tests the product Discusses the process and product Clearly explains the results Explains the possible applications

Student Self-Evaluation Checklist

Place a check mark in the column that best describes your performance.

	Did My Best	Could Improve
1. Did I listen carefully to the reading of the design brief?	❏	❏
2. Did I encourage other team members?	❏	❏
3. Did I praise the contributions of other team members?	❏	❏
4. Did I use my time wisely?	❏	❏
5. Did I consider the needs of others by speaking quietly?	❏	❏
6. Did I contribute to discussions?	❏	❏
7. Did I do everything that was expected of me in my assigned role?	❏	❏
8. Did I do my share of the work?	❏	❏
9. Did I work cooperatively with others?	❏	❏
10. Did I take pride in doing my best?	❏	❏
11. Did I continue even when faced with problems?	❏	❏
12. Did I consider the effects of my work on the environment?	❏	❏

KidTech © Dale Seymour Publications®

Design Brief Recording Sheet

Team Members Jason, Stephanie, Tom, Jill

Clarifying the Problem

What do we have to do?

We have to make a treasure box with an attached lid.

Standards we must meet

It should be attractively decorated. Sides of the box cannot measure more than 5 inches. It will be used by a 5-year-old.

Generating Ideas for Possible Solutions to the Problem

Sketches

Things to find out or materials we'll need

What do 5-year-olds like? What will we use to attach the lid? How will we finish the edges so that they look neat? Should we put something on it to keep it closed? What will be the best adhesive for attaching the cardboard panels to the wood?

Solving the Problem

Tasks	Assigned to	Comments by team members
Survey 5-year-olds to find out what they would like on the outside of the box.	Jason	Jason made up a survey and conducted it with the 5-year-olds. They wanted letters on the box and they wanted bright colors.
Find ways to attach the lid	Stephanie	Stephanie found that hinges made of metal or cardboard could be used. We decided to use cardboard.
Find something in the classoom that we can use to finish the edges of the box.	Tom	Tom found some colored tape.
Find a way to keep the box closed.	Jill	Jill discovered that we could use two push pins with ribbon, ribbon stapled to the wood, Velcro® tabs, or a latch to keep the box closed. We decided to staple ribbon to the wood.

Solving the Problem

Tasks	Assigned to	Comments by team members
Find out which adhesive will be best for attaching the cardboard panels to the wood.	Everybody	We experimented with this when we were ready to attach the panels. The glue gun worked best.
Make three 5-inch squares and four $4\frac{1}{4}$-inch strips.	Everybody	Jill and Tom each made one, and Stephanie and I each made two. Stephanie and Tom cut the strips.
Build the squares into a box.	Stephanie and Tom	They did a good job.
Make the cardboard panels and decorate them.	Jason and Jill	These took longer than we thought, so Stephanie and I helped them after we finished the box frame.
Attach the panels to the frame.	Everybody	We had to be careful to glue them exactly so that the edges were neat. Everybody helped to get this done.
Attach the lid.	Stephanie and Jill	They did this very neatly. They cut a cardboard hinge that was the same length as the cover and the top of the box.
Staple the ribbon to the wood frame.	Tom	This was hard to do, so we changed it. We glued the ribbon on with the glue gun.
Finish the edges.	Everybody	This looks great!

Test the Solution and Discuss the Results

Trials	Discussion
When we finished our box, we showed it to six 5-year-olds to see if they liked it. They all did.	Doing a survey with the 5-year-olds was a good idea because then we knew what they wanted.

What modifications were needed?   None

What improvements could be made?   We like it the way it is.

How could our team have performed better?

We worked well together. Everybody shared the work and helped each other.

What changes would we make for next time?   None. We were happy with our results.

Sample Journal Pages from Tom

Day 1
Today we got our new design brief and started to complete our recording sheet. We made the sketches for the box. It will be made of two 5- by 5-inch squares joined by $4\frac{1}{4}$-inch pieces. The cover will be another 5- by 5-inch square. This will make it no more than 5 inches on every side. It took us awhile to figure out the measurements. We almost forgot to allow for the height of the cover when we were deciding how high the box should be.

Day 2
Today we talked about how we will decorate the outside. We talked about drawing animals, letters, fairy tale, or cartoon characters on the cardboard sides. We were not sure what five-year-olds would like. We decided to do a survey to find out. We made a list of all the things we have to do and divided them up. This was the hardest part because we had to list everything we were going to do in the right order. We wanted to start building right away, but we always have to do this first.

Day 3
We have to find out what the children like before we can start the sides, so Jason made up the survey and asked the teacher if he could do it with the 5-year-olds. She said she has to ask their teacher first. While Jason was doing this, Stephanie, Jill, and I worked on other things on our "To Do" list. I found some shiny blue tape to finish the edges of the box. The principal was using it to make letters for labeling bins. Stephanie found out about hinges. She looked in a catalogue. There are different sizes and kinds of metal hinges. She said we could also use cardboard the way we did when we made our dump trucks. We could cut strips from old colored file folders. We decided to use the cardboard because it would match the box. Jill looked for a way to keep the box closed. She said that we could use two push pins with ribbon, ribbon stapled to the wood, Velcro® tabs, or a latch. We decided to staple ribbon to the wood.

Day 4
Today we made our squares. Jill and I each made one, and Stephanie and Jason each made two. We used all 45-degree angle cuts. Jason finished the survey, so he and Jill started to decorate the cardboard panels. They are using bright capital letters so that the box will look like a big block. Stephanie and I cut the pieces to join the squares. It was a good day!

Day 5

Stephanie and I finished the box frame. Jill and Jason were still working on the panels, so we had to help them. They took longer than we thought because you had to draw everything first and then color them. We made six panels. We were glad to get them done, because it was tedious work.

Day 6

Today we glued the panels onto the wood frame. We tried white glue first, but it was not sticking too well on the wood. We finally decided on the glue gun. We had to be very careful because the glue dried fast. We had to get them just perfect so that the box would be neat. We also put the hinge on. We used a strip from a blue file folder and cut it the same length as the cover. All we have to do now is attach the ribbon. The box looks really good. I think you could sell it at a craft show!

Day 7

We had some trouble today. We could not get the ribbon stapled to the wood. It was hard to hold it in place and push hard enough for the staple to go in. There was no way for us to hold it in place to staple it. It would have been easier if we had done this before we put the frame together because we could have held the wood flat on the desk. We decided to glue it around the frame. We had to take the cardboard panel off at the top so we could get the ribbon around the wood to glue it. Then we had to glue the panel back in place. It turned out fine, but we should have thought of this sooner. All we have to do now is finish the edges.

Day 8

Today we finished our product. All we had to do was put the tape on to finish the edges. It was difficult because we had to roll out the tape, hold it against the box, and then cut it to get the exact size. I guess we could have measured, but this seemed just as easy. It was also important to try to line up the tape. On the self-evaluation checklist, I marked that I did my best on each item.

Teacher Evaluation

The completed Design Brief Recording Sheet, journal pages, and Student Self-Evaluation Checklist show exemplary performance for this group. This record of their work along with observational data collected by the teacher, the treasure box itself, and their presentation of the product show that these students have satisfied the criteria for quality work.

TEACHER NOTES FOR PRACTICE DESIGN BRIEF

The practice design brief is intended to provide a simple problem that will allow students to focus on the design process, the roles and responsibilities of team members, and the assessment techniques that will be used. It will also give them an opportunity to learn some basic construction techniques. Work through this first design brief slowly, always referring to the listed roles and responsibilities of each team member. Encourage students to share and compare ideas throughout the process.

This design brief requires students to make a 15-centimeter square frame. Students must carefully consider what they are going to do before they start cutting the wood. They have to think about the length of the pieces they must cut and also about what kind of cuts they are going to make.

They can make straight or angled cuts, but of course their choice will make a difference in the final measurements. Usually their first idea is to make straight cuts and join four pieces measuring 15 centimeters each to form the square. For this reason, it is wise for them to design their solution first. They should begin with a simple sketch of what their square will look like, including measurements. This will lead them to think about lengths and end cuts. Their sketch should show three-dimensional wooden strips, not just one-dimensional lines. For some students, it will help if they can make a model using centimeter cubes. When they can explain their design, they are ready to begin cutting. They need to know how to use a bench hook or miter box, how to use a saw, and how to join wooden strips, using triangle corners. These skills are addressed in the Tools section on page 4 and in the construction section on page 30.

Possible Solution

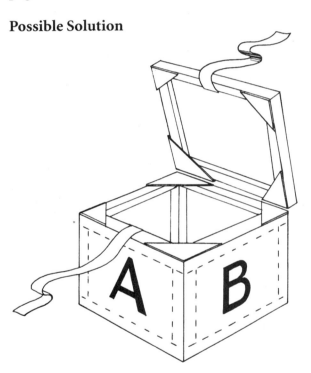

Practice Design Brief

You are about to learn about technology, how it affects your lives, and how it can be used to provide for your needs. Using technology means using what you know, with the tools you have and the materials you can gather, to produce what you need. One of the most important skills you can learn is how to solve a problem or accomplish a task with a limited supply of materials or within very specific guidelines. It is also important to be able to do this with others.

Each design brief will give you an opportunity to work on a team to solve a problem, using technology. Working well as a team member is just as important as solving the problem. Each member of the team will have a different role to play and be given very specific tasks to do. As you work through several design briefs, you will have a chance to perform each of the roles. In order to perform well on a team, team members must understand their roles and support one another in performing their tasks.

Your first task is to design a square frame that measures 15 centimeters on each side, and then construct it, using strips of wood.

BASIC TECHNIQUES

To complete the design briefs, students require some basic knowledge. That knowledge varies with the design brief, but generally falls into three categories: they will need to know basic construction techniques; sometimes they will want to know how to make their creation move; and once they require movement, they need a power source. The Basic Techniques section outlines knowledge for construction, movement, and power and gives suggestions on how to teach these concepts. Individual design briefs will refer you to the Basic Techniques section to indicate what students should know before they can successfully complete the brief.

CONSTRUCTION

Lightweight wood in 10-mm-square strips is easy for students to work with when building structures or models. Using heavy, thick wood is frustrating without power tools. Use cardboard corners or

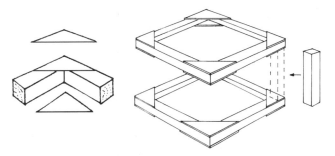

strips to join wood together quickly and easily. Use cardboard strips inside corners or cross strips to make joints stronger. White glue, carpenter's glue, and glue applied with a hot-glue gun are the most effective adhesives for construction.

Plastic straws or paper art straws are also an inexpensive material for building structures. They are easy to cut and can also be joined using cardboard strips or corners. They can also be bent and joined together with thinner straws, pipe cleaners, paper fasteners, glue, or wooden stir sticks to form irregular shapes easily. Straws also make good axles for cardboard-box vehicles.

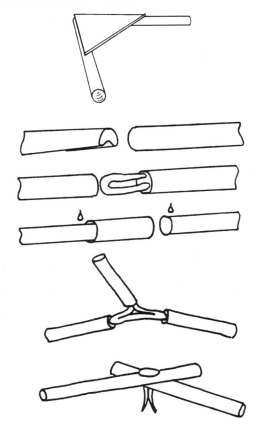

Cardboard boxes of all shapes and sizes are an invaluable resource. Many boxes can be taken apart and then reassembled inside out. Students should examine how the box has been put together, then gently undo the three seams, and lay the box out flat.

Cutting windows, doors, and holes for axles, as well as painting, can be done while the box is apart. To reassemble the box, students fold the seams the opposite way. Boxes can be reassembled temporarily, using paper clips, sticky putty, or tape so that students can assess their progress and then make revisions in their designs. Once they are happy with it, they can glue the box together. Basic shapes can be joined together to make more elaborate structures.

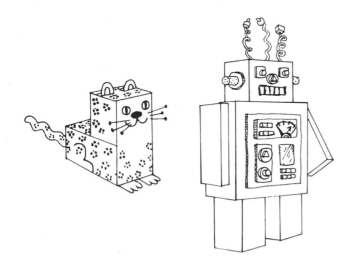

MOVEMENT

A variety of techniques can be used to create movement.

Wheels

Wheels for vehicles can be purchased or can be made from many materials such as film container covers, buttons, pop tops, jar lids, thread spools, ribbon spools, and paper plates. They can be mounted on axles and attached to vehicles in a variety of ways. Plastic caps make good wheels for cardboard or other lightweight vehicles. They can be mounted on axles by gluing a plastic craft bead or a small piece of plastic or rubber tubing onto the wheel and then inserting a plastic straw into the hole of the bead and gluing it into place. Card-

board triangles with a hole punched at one point can be used to attach the axle to the vehicle.

Making cardboard wheels with wooden hubs

1. Use a compass or trace around a circular shape to draw the wheels on heavy cardboard. (2 to 4 inches is the recommended size for most applications.)

2. Make hubs for the wheels out of small squares of wood. (Save scraps for doing things like this.) Mark the center of the hub by drawing lines from corner to corner. Drill a hole in the center of the hub that will fit the axle being used.

3. Find the center of the wheel by tracing the wheel on paper, cutting out its shape, and then folding the paper in half twice in opposite directions. After marking the center of the paper, place the paper on top of the wheel and mark the center of the wheel with a compass, pin, nail, or heavy marker.

4. Glue the hub onto the center of the wheel. Match the center of the hub to the center of the wheel.

5. Measure the width of the vehicle to determine the length of the axle you will need. Cut dowels, wooden skewers, lollipop sticks, or straws for the axles.

6. When the vehicle is finished, glue the axle to the center of the hub. Wheel and axle will turn as one unit.

Making hubs from strips of cardboard

1. Cut 12 strips of cardboard (3 for each wheel) about $2\frac{1}{2}$ inches by $\frac{1}{4}$ inch.

2. Fold each strip in half and glue them back to back leaving space for the axle in the center.

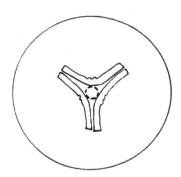

3. Follow the same steps 4–6 as you did when making a wooden hub.

Plastic tubing can be used instead of hubs to secure axle to wheels while leaving them free to turn. Make a hole in the wheel big enough for the axle, but too small for the outer diameter of plastic tubing that fits on the axle. Glue a piece of plastic tubing to the axle on each side of the wheel. Secure the axle on the outside of the wheel.

Making large wheels from paper plates

1. To make 2 wheels, mark the middle of 4 plates inside and outside.

2. After drilling the axle hole in the hub, glue each hub onto the center of a plate.

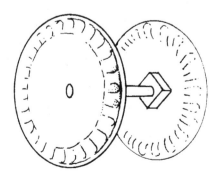

3. Glue two pairs of paper plates face to face with the hub between each pair of plates.

4. Use a compass to put a hole through the center of one plate in each pair for the axle so that it can be glued into the hub.

Securing wheels to vehicles

If the wheel will turn independently of the axle, the axle can be glued directly to the vehicle. If the wheels and axle form one unit, the axles can go through holes drilled in the frame of the vehicle or through triangles attached to the vehicle frame.

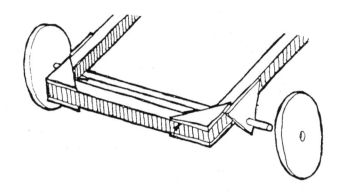

Pulleys

A pulley is like a wheel with a groove around its circumference. It can be used with a string or chain to transfer energy from one place to another. There are a variety of pulley systems that can be used.

Fixed Pulley System

The fixed pulley system (see page 31) uses a wheel and rope. The grooved rim of the wheel holds the rope. When the rope is pulled, it turns the wheel. Pulling one end of the rope lifts a load attached to the other end. A fixed pulley does not make it possible to do things faster, nor does it save effort. It changes the direction of the effort. Since it is easier to pull something down than lift it up, the job of raising an object is made easier.

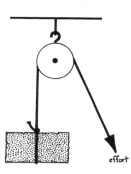

The Movable Pulley System

The movable pulley system enables people to lift many times their weight. It not only can change the direction of effort, but it saves effort. A movable pulley system involves at least two pulleys, one being movable. This pulley is attached to the load, and the other is fixed and acts as the support. The rope is attached to the fixed pulley; it runs down and around the movable pulley, then around the fixed pulley. The movable pulley moves with the load as the rope is pulled and the load is lifted.

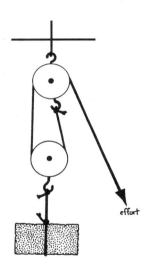

Pulleys can be purchased or can be made from thread or ribbon spools. Try the following activities with your class as an introduction to pulleys.

♦ **Materials**   plastic grocery bag, textbook, thread or ribbon spool, spring scale, piece of wire, ruler

1. Have students place the book in the grocery bag and then lift the bag, using the spring scale. Record the effort used.

2. Make a pulley by putting the wire through the spool and joining the ends of the wire into a hook.

3. Span two desks with the ruler. Hold or weigh the ends of the ruler down.

4. Hang the pulley on the ruler.

5. Put the bag on the floor between the two desks, tie a string to the bag, and bring the string over the spool.

6. Have students lift the bag, using the pulley. Ask, Was it more or less difficult, or as difficult as when it was lifted with the spring scale? (It should seem easier because the direction of effort is down instead of up.)

7. Tie the end of the string to the spring scale and lift the bag again, using the pulley. Measure the effort. Ask, Was the effort much less? (It should not be; a fixed pulley does not save effort but merely makes it easier to lift something by changing the direction of the effort. It is easier to pull down than lift up.)

Gears

Gears are just like wheels with teeth. The teeth of one gear fit into the teeth of another gear, so one gear turns another. Gears are used in many machines to slow down or speed up movement or to transfer movement from one part of a machine to another. Gears can be made from any type of disk, but the design must be followed very precisely. You will need two disks for each gear. They can be made from heavy cardboard the same way wheels are made.

1. Mark the center of each disk and put a hole in it.

2. Center the disk on the 6- or 8-point gear-making guide provided on page 32. Use a ruler to draw the 6- or 8-point gear guide lines onto the disk.

3. Cut 6 or 8 equal lengths of wooden dowel, or use wooden stir sticks. The length must be greater than the radius of the disk. Glue them from the edge of the center hole along the gear guide lines. Scraps of wood can be glued alongside the dowel or stir sticks to make stronger gears.

4. Cover the disk and dowel or stir sticks with the second disk.

The gears can be mounted to a vehicle, structure, or machine. To change the direction of rotation from one gear to another, use two gears that are the same size.

To keep the direction the same, use three gears that are the same size.

To decrease speed and change direction, go from a smaller gear to a larger gear.

To increase speed and change direction, go from a larger gear to a smaller gear.

Gear-Making Guides

6-Point Guide

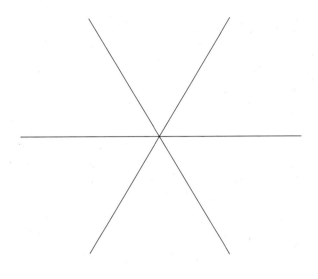

8-Point Guide

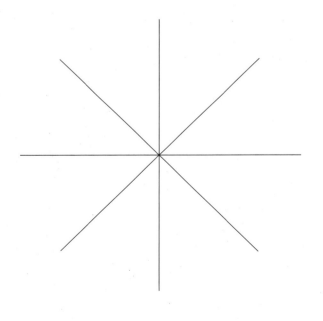

Try the following activities with the class as an introduction to gears.

♦ **Materials**   metal bottle caps with a nail hole in the center of each; finishing or carpet nails, or paper fasteners; a thick piece of Styrofoam

1. Have students attach one bottle cap to the Styrofoam.

2. Attach a second bottle cap so that its grooves fit into the grooves of the first one.

3. Have students turn the first cap. The second cap should turn. (If it doesn't, students have to figure out the importance of accurately lining up the grooves or having the caps fastened securely enough.)

4. Note which direction the second cap turns when the first one is turned. (It should turn in the opposite direction.)

5. Have students attach a third cap so that its grooves fit into the grooves of the second one.

6. Note the direction in which the third cap turns when the first one is turned. (It should turn in the same direction.)

7. Have students make gears of different sizes. Let them repeat the activities with gears of different sizes to discover that larger gears will make smaller gears turn faster because the larger gear has a greater circumference. The smaller gear has to make several turns for every revolution of the larger gear.

Drive Belts

Drive belts work as gears do by transferring movement from one wheel to another. Drive belts are easier to work with and simple to make. Two grooved wheels (like those used for pulleys) with axles can be joined with a rubber band. The speed of the pulley wheels turning depends on the relative size of the two connected wheels just as size

affects the relative speed of gears. The larger wheel will rotate more slowly than the smaller wheel. The greater the difference in size between the two wheels, the greater the difference in speed. The wheels will turn in the same direction when joined like this.

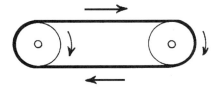

The wheels will turn in opposite directions when joined like this.

Speed can be controlled by using wheels of different sizes.

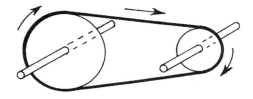

Try the following activities as an introduction to drive belts.

1. Attach a thread spool to a piece of wood or a corkboard with a nail so that it turns freely. Attach a larger ribbon spool 6 inches away. Put a rubber band around both spools so that when one is turned, the other will turn. Use a dark marker to put a vertical line on each spool. Have students count how many times the smaller spool turns for a single rotation of the larger spool. (The smaller spool will make more than one full rotation and will seem to turn faster. Both spools will turn in the same direction.)

2. Now have students investigate whether the drive belt design influences the rotation of the spools. Move the smaller spool 2 inches farther away. Once again have students count how many times the smaller spool turns as the larger spool rotates once. (There should be no change.)

3. Twist the rubber band so that it forms a cross between the spools. Have students observe the small spool as the large spool rotates once again. (The small spool will make more than one full rotation, it will turn faster, and it will turn in the opposite direction.)

4. Have students explore using different size spools, different drive belts, and different numbers of spools. They can also look for familiar machines that use drive belts, such as bicycle chains, clothesline pulleys, and some motors.

POWER

Pushing or pulling a vehicle is a simple means of creating movement. The student provides the energy.

Inclined Plane Power

Students may wish to investigate how an inclined plane can be used to power a vehicle. This will be especially appropriate if the students are studying simple machines. The weight of a vehicle on an inclined plane moves it down the ramp. People power must be used again every time the vehicle stops. If you want the vehicle to end up in its original position, you'll have to turn the ramp around and move it where the vehicle has stopped. This is also a good time for students to investigate friction and how surfaces affect speed. Try changing the angle of the ramp and using different surfaces.

Air Power

Attaching a sail to a vehicle that can then be powered by the wind, a fan, or even the student's breath is another simple method of providing movement. This would be a good power source to explore when students are studying weather or air. The sail can be made of paper or cloth and attached to a stick anchored in a hole in wood or a mound of clay.

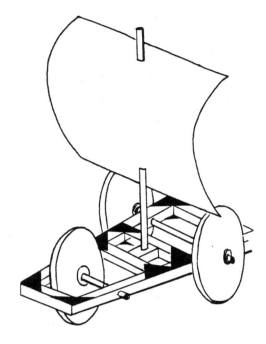

Balloon Power

The kinetic energy from the lungs can be compressed and stored in a balloon as potential energy when the balloon is sealed. The inflated balloon can be used to power a very light vehicle or support a light structure. The balloon should be secured

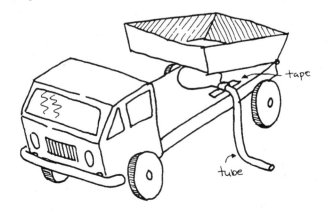

with tape. When the air from the balloon is released, it will cause the vehicle to move in the opposite direction or the structure resting above the balloon to move downward.

Propeller Power

Propellers can also be attached to a vehicle as a source of power. Propeller blades are made thin and angled so that they can cut through the air. When the propeller turns, it pushes back air and so moves forward. Something is still needed to power the propeller. This can be done by attaching the propeller to a pull cord or a rubber band. This student drawing shows a propeller with a rubber band.

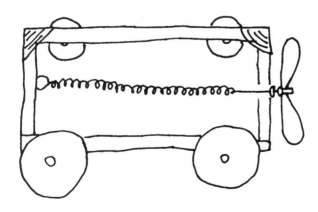

Pneumatics

Pneumatics is the use of air to make something move. It is an inexpensive and environmentally friendly source of power. Plastic tubing and syringes without needles, or balloons, allow students to harness pneumatic power that can be used to make their creations move.

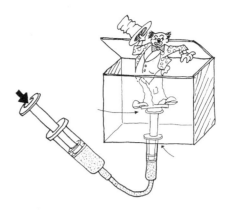

Students need opportunities to explore this source of power. Point out to your students that the syringes do not have needles. Caution them about the dangers of picking up or handling syringes if they ever see them outside the classroom. Try the following activities with your class as an introduction to pneumatic power.

1. Insert plastic tubing into a balloon and attach it with masking tape. Have students try to move light objects by blowing up the balloon through the tubing. (For hygienic reasons, individual pieces of tubing should be available for each student.)

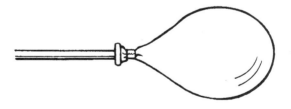

2. Attach two 10-ml syringes with a 3-cm piece of tubing. (Have the plunger of one syringe extended and the other depressed.) Have students observe what happens when they depress the extended plunger. (The depressed plunger will now extend.) Have them do this several times and discuss what is happening and why. (Depressing the plunger on the master syringe causes air to push the plunger

out on the slave syringe, or syringes if several are joined. The master syringe must always be as large or larger than the slave syringe, or syringes if several are joined.)

3. Try the activity again, using the 3-cm tubing joined to 6-cm tubing with a straight connector.

4. Try it again, using 50 cm of tubing.

5. Now try using a 20-ml syringe and two 10-ml syringes joined with a T connector.

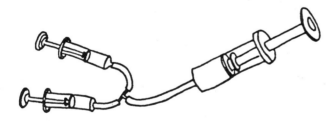

Bring only syringes without needles into the classroom and caution students to never touch used syringes they find. Once students have seen how the syringes or the balloons can be used to cause movement, they can apply what they have learned to solving a simple problem. For example, students can use pneumatic materials to make this dog come out of his house.

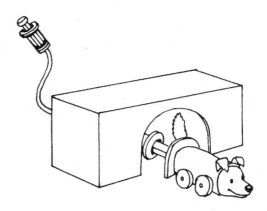

Hydraulic Power

With hydraulic power, syringes and tubing are used as in pneumatics, but they are filled with water. This is done by submersing the whole system in water, so that the tubing and syringes are filled, and water instead of air creates the power. Hydraulic power is more forceful than pneumatic power.

Rubber Band Power

Rubber bands can also be used to power vehicles without a propeller. The rubber band can be attached to the front of the vehicle's frame with a paper fastener and then wrapped around the back axle. Wind the rubber band around the axle by holding the vehicle and turning the axle several times. When the vehicle is set down and the axle released, the rubber band unwinds and turns the wheels, propelling the vehicle forward.

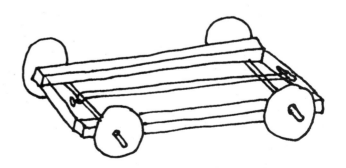

Electric Power

With a basic knowledge of electricity and electrical circuits, students can use this source of power for vehicles, switches, buzzers, and lights. The following activities can be used to introduce students to electricity and electrical circuits.

Simple Circuits

Provide each team with a D-cell battery, a low voltage flashlight bulb, and about 12 inches of wire that has bare ends of about $\frac{3}{4}$ inches. Students try to light the bulb. Once they have succeeded, have them draw and compare their solutions. Have them discuss what the solutions have in common. The following solutions are possible.

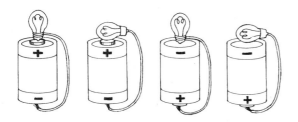

Try using two batteries or three. (Do not use more than three.) Continue the investigation using two wires with the battery and one or more bulbs.

Scientific Explanation: The battery is a dry cell. It is made up of a zinc case with a carbon rod down the center. The case is filled with chemicals that are in the form of a black paste. When a circuit is created, a chemical reaction occurs. It causes electrons to flow from the negative pole of the battery, where there are too many electrons, through the external wire to the positive pole of the battery. The chemical energy has been changed into electricity that flows along the wire conductor. The more bulbs that are in the circuit, the more resistance the electricity meets and the dimmer the light will be. If one bulb does not work or if the circuit is broken anywhere, none of the bulbs will light. You may wish to have students take a dry cell apart to examine its contents. This can be done with pliers and a screw driver. The materials inside the dry cell will not harm the skin but can stain clothing.

A Circuit Tester

Once students understand how a simple circuit works, have them experiment with different materials to discover which ones are good conductors. For this activity, each team will need a bulb holder, a D cell, a cell holder, some insulated copper wire, and a collection of metal and non-metal objects such as nails, coins, foil, tin cans, paper clips, straws, chalk, garbage bag ties, pens, pencils, erasers, and so on. Have them arrange them as shown below. Place the material to be tested between the two wires. Materials that are good conductors will complete the circuit and light the bulb.

Parallel Circuits

When students are comfortable with circuits in series, you may wish to have them investigate parallel circuits. Some of the students may have already created parallel circuits by accident when they were experimenting with two bulbs and several wires. They may have noticed that when the circuit is set up a certain way, one bulb can be burned out and the other in the circuit will still work, or that even when they added bulbs, they all remained bright. If this occurred, the students had created a parallel circuit in which each bulb was in its own circuit.

Following is an example of a parallel circuit.

Note: A knowledge of both types of circuits will be useful to students in their designs.

There will be times when a parallel circuit will be most effective—for example, headlights on a vehicle that must be as bright as possible but must also be reliable, so that if one light burns out, the other one would still work. A series circuit would be more efficient to power an energy-saving device where brightness was not required.

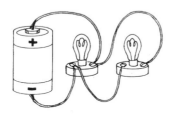

Hidden Circuits

Circuit testers can also be used to locate hidden circuits. As an assessment of students' understanding of circuits, prepare a hidden circuit board for the students to solve. It can be made simply by using paper fasteners, with one or more pairs connected with insulated wire, attached to the inside of a box, or sandwiched between two pieces of cardboard that are then stapled together. Label the fasteners with numbers or letters. Students can predict and then use a circuit tester to identify the hidden circuits. Students can also make their own hidden circuit boards for classmates to solve.

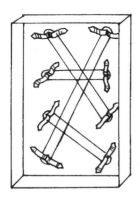 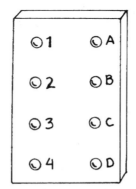

Switches

Simple switches can be made so that students can control the flow of electricity in a circuit. Switches can be made from strips of cardboard wrapped in aluminum foil or from a strip of copper or aluminum. It should be assembled as shown in the diagram. Challenge students to find a way to use two switches to turn off a light from different locations.

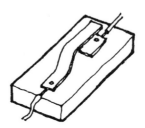

Motors

Electric motors can be connected to a battery and controlled by a switch. This can be useful in the construction of machines or vehicles. In order to slow the motor down and increase the power provided, use the motors in combination with a drive belt. Commercial kits such as Lego® Dacta can be used to provide power. Students can also make a prototype of their solution with Lego® to test the design, and then create their final product.

TEACHER NOTES FOR DESIGN BRIEFS

DESIGN BRIEFS CAN BE USED TO ADDRESS A WIDE variety of cross-curricular knowledge, skills, and values. Focus on those that meet your needs. The teacher material for each design brief has suggestions for extensions to other areas of the curriculum; you may find other good extensions. With your students, establish criteria for quality work in each of these areas based on the standards of performance in your school.

The extent to which you are addressing other curriculum areas should be considered when deciding how much time should be spent on a design brief. Generally students will require the following time allotments to work through the design process.

1. Clarify the problem — 45 minutes to 1 hour

2. Generate ideas for a possible solution—1 hour

3. Implement a solution to the problem—2 to 4 hours

4. Test the solution and discuss the results— 2 hours

Before assigning any of the design briefs, it is important that teachers have a place for integrating design technology into their curriculum.

COMMUNICATION DESIGN BRIEFS

The communication briefs involve market analysis and product promotion. Students will form companies and decide what people want or need, how to let people know it is available, and how to encourage them to buy it. The design briefs will require students to conduct surveys, design product names and logos, and prepare advertising campaigns for a variety of media. These design briefs are an excellent vehicle for introducing students to design technology. They could also be used throughout the year as a cross-curricular extension to other design briefs. After students build a product, they could consider how it could be marketed.

Communication Design Brief 1
What's in a Name?

Begin this brief by having students collect as many logos as possible. Have them divide the logos into categories.

- logos that display only the company name

- logos that say or demonstrate what the company does

- logos that say or demonstrate information about the company's values

Discuss why each logo is successful. Encourage students to look at any images used, the lettering, border, shape, background, and colors. If possible, reproduce the finished logos so that each group can display their company logo on a shirt or hat.

♦ **Cross-Curricular Extensions** Have students make company stationary and business cards. Have them do a poster display of the logos they collect. Students should determine the cost of reproducing their logos.

♦ **Materials**   art supplies, newspapers, magazines

Communication Design Brief 2
Name That Product

An important aspect of this design brief is product name analysis. Students should begin by going through newspapers and magazines and collecting pictures of many different products. Have students group product names under these headings.

- names that focus on a catchy sound

- names that focus on cost

- names that focus on uniqueness

- names that focus on what the product does

- names that focus on being environmentally friendly

- names that focus on luxury

- names that focus on safety

- names that focus on the target consumer

This will give students some ideas about naming their products. Ask, How could you find out which name probably would be the most successful if you were actually to market the product? Students could test-market their names by surveying another class.

♦ **Cross-Curricular Extensions**   Have students illustrate the results of their survey or product name analysis on a graph. Discuss the probability that girls or boys of a certain age group might pick one name over another. Consider how successful the product might be in another country.

♦ **Materials**   art supplies, newspapers, magazines

Communication Design Brief 3
Print Ad Campaign

Students must do a critical analysis of a variety of print ads so that they can make an informed decision about the media that will work best for the product they chose. Once they decide on the type of print ad they will produce, students must determine which aspects of their product to promote. The approach they take should be consistent with the product name they have chosen. Promotion will have to focus on the company image as suggested by the name.

♦ **Cross-Curricular Extensions**   Discuss the power of advertising and its impact on children. Have students research government regulations on advertising. Ask, Are there any products that cannot be advertised? Look at ad campaigns that are aimed at educating the public about an issue, such as safety, the environment, or political candidates.

♦ **Materials**   art supplies, newspapers, magazines, TV

Communication Design Brief 4
Don't Change That Channel

In preparation for this design brief, students should have opportunities to study as many radio and television ads as possible. They should look at what makes them different from print ads. Direct their attention to the sound effects, the emotions aroused in the viewer, the choice of actors or characters and their appeal, the amount of information given, and so on. After this exercise, students should be better prepared to discuss the features they would like to include in their own ads. Students should also have had some practice in preparing scripts, acting, recording, and videotaping their own work.

♦ **Cross-Curricular Extensions**   Have students research the cost of air time at local radio and

TV stations. Have students prepare a budget and schedule for the ads they would like to run.

♦ **Materials** tape recorder or video camera, radio or TV

Communication Design Brief 5
Follow That Ad!

These moving ads can be simple pull tabs containing the commercial message. The message is written on a strip and then joined to make a circular message that keeps reappearing as you pull the different frames past a window. You may wish to make this a mechanical challenge for some students. The challenge for students is how to write the message so that pertinent information is revealed a little at a time. Each frame must build the suspense and encourage the consumer to keep watching. It would be excellent if students had an opportunity to see moving ads in their community. You could videotape some and look at what makes them effective.

♦ **Cross-Curricular Extensions** Have students create a survey for people who view their ad to learn their reaction to the ad. They should find out the cost of having a message run on a moving ad.

♦ **Materials** mural paper, poster board, tape

Communication Design Brief 6
Market Analysis

This design brief gives students the opportunity to take an organized look at a realistic market situation. They should begin to realize that the choice of models built by each manufacturer is no accident. They are appealing to a particular market. Encourage them to identify any gaps that exist. Look at the gaps in view of your community. Ask, What are the needs of your community? Are there many large families? What is the income range? Do families have one or two cars? What are the average ages of children? Will there be many teenagers looking for inexpensive cars?

♦ **Cross-Curricular Extensions** Have students conduct surveys in their own homes about what family members look for when they buy a car. Are they happy with the car dealerships in their community? Why? Why not? Which car gives best value for money? Look at the resale value of cars and discuss depreciation.

♦ **Materials** newspapers, car magazines

Communication Design Brief 7
What Does It Cost?

Students should figure out very quickly that their profit will be greater if they limit the number of choices they offer. This way they can buy in bulk, cut down on the kinds of equipment they need, and strive for quality as opposed to variety. Actually having the opportunity to organize such an activity provides an authentic learning experience for students. The figures will be meaningful for them. The students should be expected to keep very accurate records, including receipts.

♦ **Cross-Curricular Extensions** Have students advertise their event. Let the class decide how they will spend any profit made. Discuss the concept of putting profits back into a business.

♦ **Materials** These will vary depending on student choices of crafts.

Communication Design Brief 8
Consumer Survey

To be successful with this design brief, students should have had some experience writing survey questions and conducting surveys. You may want to spend some time practicing this. The bookmaking itself should be part of your regular language lessons. Students should have an opportunity to see many unusual kinds of children's books, including poster books, fabric-covered books, accordion books, books with pocket messages, flaps, ribbons, or padded covers, books that are shaped differently, and books with three-dimensional features. Students should be striving for word-perfect copy with their best illustrations. The results of their survey should be evident in the book they create.

♦ **Cross-Curricular Extensions** Have students sell their books at a book fair that they organize and advertise.

Communication Design Brief 9
Promotional Flyer

Have students bring in a selection of promotional flyers received in their home mail.

Approach local book stores for flyers. Try to get an author or a publisher to come in to talk to the students. Decisions on where they will sell their book and send their flyer depend on the audience they are targeting. After they have had an opportunity to discuss and analyze this information, they should be ready to make their own flyer. This can be a regular language and art lesson at this point, with the same criteria you would set for any written assignment.

♦ **Cross-Curricular Extensions** Have students organize their books for a book club. They would group them according to topic and decide on a price. They could also design an order form and a name for their club.

♦ **Materials** sample flyers

Communication Design Brief 10
Does the End Justify the Means?

In this design brief, students look at their responsibility to use technology wisely.

In their presentation, they should discuss some of the following issues.

- wildlife—The over-development of land results in loss of animal habitats.

- greenhouse effect—Gases such as methane, CFCs, nitrogen oxide, and carbon dioxide are gathering in the atmosphere, destroying the ozone layer. This exposes us to more of the sun's dangerous rays.

- automated farming—More food is produced, but jobs are lost.

- fast foods—They not often not healthy and their packaging is filling our landfill sites. The demand for meat has resulted in clearing forests to raise more animals. Time is saved in food preparation but less time may be spent eating together as a family.

- use of robots—People are freed from monotonous and sometimes dangerous work, but jobs are lost.

- natural resources—They are in finite supply, so recycling is important.

- pollution—Industry often has harmful effects on water, animals, soil, and the atmosphere.

- rain forests—Local inhabitants need to support their families, but animal habitats are lost and soil erodes when rain forests are destroyed.

- landfill sites—We may be producing more garbage than we can handle. We need to find ways to reduce waste.

- use of energy—Conserving energy is important. We need to find ways to conserve more.

- organic farming—The use of chemicals to ward off disease and increase the size of crops affects humans, animals, and soil.

- developing countries—Industry improves the standard of living, but often at the expense of the environment.

♦ **Cross-Curricular Extensions**   Have students take some action, such as writing to politicians to express their views, getting more information from environmental groups, and looking at what they can do to make a difference.

MACHINES DESIGN BRIEFS

The machines design briefs can be used as the foundation for an integrated unit on simple machines or inventions or to complement an existing unit.

Machines Design Brief 1
Time Is Up

This brief requires students to make a simple see-saw timer that has a paper cup filled with sand on each end. One of the cups should have a tiny pin hole in it that will allow the sand to drain out in 15 minutes. When it does, the lighter end of the see-saw will go up. The real problem lies in finding out how much sand to use, where to put the hole, and how big the hole should be in order to make the see-saw go up in the required time. This must be done through trial and error. Once it is done, students must figure out that the level of sand in the leaky cup must be marked, so that students can refill it each time to start the timer again. They must also figure out a way to block the hole until they want the timer to be activated. A piece of tape or a push tack are two possibilities.

♦ **Cross-Curricular Extensions**   Work on time to the minute or second. Have students fill different cups for different amounts of time so that the timer could be used to time different events. Students can make charts of the relationship between the amount of sand and the time it takes to go through the hole. Have them experiment with other materials such as rice or salt and look at time pieces from different cultures.

♦ **Materials**   block of wood, two paper cups, tape, ruler or flat piece of wood, sand

♦ **Possible Solution**

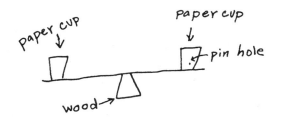

Machines Design Brief 2
Save That Kitty

To complete this design brief, students need to have had some experience with making and using a movable pulley system. Background information and preparatory activities on movement are on pages 29–31. Students will first have to make a carrier large enough to hold the animal. When setting up the pulley system, students must attach the movable pulley to the load—in this case, the animal in the carrier. Emphasize that this is a make-believe situation

♦ **Cross-Curricular Extensions**   Have students weigh their animals and find the heaviest load that can be lifted. Have them write possible explanations of how the cat came to be on the ledge. Ask, What responsibility do we have when we discover any living thing in trouble? Talk about the safety factors involved. Ask if they should have risked going down for the cat and if the cat could have been dangerous.

♦ **Materials**   string, thread or ribbon spools, cloth

Machines Design Brief 3
A Changing Sky

This brief requires that students have some knowledge of a belt and pulley system. Background information and introductory activities are provided on pags 30–33. They must begin by making their backdrop. The brief requires that it be attractive, so students must take care when doing their initial art work. Once this is completed, the next task is to set up the belt and pulley system. The system must be designed so that when the small spool is turned, it turns the large lid and causes the sun to come up and the moon to go down. Students will have to experiment with the placement of the sun and moon on the system and with the best combination of different size spools or lids to use.

- ◆ **Cross-Curricular Extensions** Have students create a play, using their backdrop, and perform it for younger students. Students may wish to investigate the work and the cost involved in putting on a large production. Ask, What kinds of careers are involved in bringing a production to an audience?

- ◆ **Materials** different size lids and spools, straws, string, rubber bands, nails, a large piece of wood or cardboard for the backdrop, art supplies

- ◆ **Possible Solution**

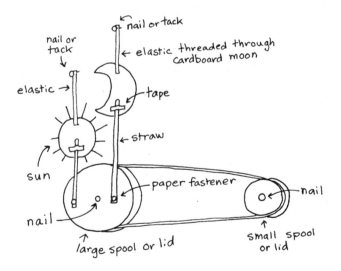

Machines Design Brief 4
Open for Business

This design brief requires that students have a knowledge of gear systems. Background information about gears and introductory activities are provided on pages 31–33. Emphasis should be on producing a working model rather than on the aesthetic quality of the shutters. If you want to make this a more elaborate project, you could have students make their lemonade stand more elaborate by decorating the outside. Rather then using a box, they could use a commercial building set or a structure built in another design brief.

- ◆ **Cross-Curricular Extensions** Have students discuss the benefits of young people being encouraged to have small businesses. What does a business teach children? (Money management, commitment, responsibility, making change, talking with people, and so on.)

- ◆ **Materials** heavy cardboard, gear tracing sheet (copy right half of page 32), dowel, saw, bench hook, glue, nails, wood for a base, straws, ice cream sticks

- ◆ **Possible Solution**

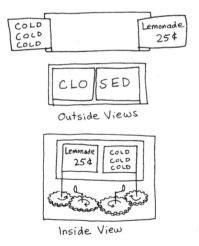

Machines Design Brief 5
Talking All Night

Students should have had some opportunities to learn about and work with pulleys in preparation for this design brief. It would be an excellent follow-up to a unit on simple machines. Basic background information and introductory activities on pulleys are provided on pages 30–31. The solution to this brief can be very simple. You may choose to make it more challenging by asking students to make a movable pulley system or a more extensive system that would involve other campers.

♦ **Cross-Curricular Extensions** Have students write the messages that the two friends might exchange. Have them devise a code in case the messages are intercepted. Imagine that friends are from two different cultures. Students should research the culture and then write a character sketch for each of the two friends.

♦ **Materials** string, clothespins, 4 spools, pieces of wood, nails

Machines Design Brief 6
Your Baby Sister's Pool

The task in this design brief is to set up a water wheel in a lake. A cutout plastic bottle or dish makes a good lake. The wheel will scoop out water as it turns and pour it down the chute. The chute should run from the lake to the kiddie pool. The only background knowledge required is how to make a wheel and attach it to an axle. Introductory activities on wheels are provided on page 29.

♦ **Cross-Curricular Extensions** Have students measure the capacity of the containers and the volume of water they carry. Have them time the filling of the pool. Discuss the uses of such a system for farmers in developing countries. Look at water systems in your area. Have students design a filtering system.

♦ **Materials** large plastic bottle or dish, small dish or lid cover, large lid covers for wheels, 6 small plastic cups or bottle tops, nails or glue gun, wooden dowel, wooden strips, large milk carton to make the chute

♦ **Possible Solution**

Machines Design Brief 7
Next Please

Students will have to use a variety of techniques to complete this design brief. A knowledge of basic construction techniques, including a knowledge of how to make a conveyor belt, is required. See the Basic Techniques section on page 28. The real challenge is to combine these techniques and create a working model. The straws or corrugated paper must line up properly, and the belt must be tight enough but not too tight to move. If students use corrugated cardboard to make the rack and pinion, it must be fastened very securely and precisely to work.

♦ **Cross-Curricular Extensions** Have students time a repetitive task, doing it on their own, then using an assembly line and a conveyor. Make a list of places where students have seen conveyor systems. Have them think about how it might be designed to move heavy objects. Discuss the impact of technology on the work force and how the needs and concerns of the workers changed with advances in technology.

♦ **Materials** 2 spools, large plastic straws or corrugated cardboard, wooden dowel that will fit into the spool, construction paper or ribbon that will fit around the spool, wooden strips, glue, corners, drill

♦ **Possible Solution**

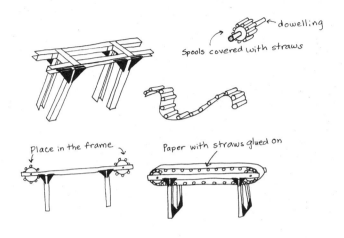

dowelling

Spools covered with straws

Place in the frame

Paper with straws glued on

Machines Design Brief 8
Pick That Up

Students can really let their imaginations run wild in this design brief. There is no background teaching required. This product will depend on what materials students are able to gather. They may then choose to make one multipurpose tool or they may design it so that it has a variety of add-on components for different purposes. Students should have the opportunity to look at a variety of commercials and print ads so that they can make an informed choice about how they want to market their product. It is important that they be given sufficient time to present their finished product with its accompanying sales pitch.

♦ **Cross-Curricular Extensions** Students should research market demand for these products. Competitive pricing, target consumers, and options for advertising and distribution should be considered. Profit after taxes is worth discussing. The increase in environmental awareness and its impact on product design, and on

packaging in particular, are important points for discussion.

♦ **Materials** long pole or stick, plastic bottles, cans, tape, combs, string, magnets, nails, twist ties

Machines Design Brief 9
Stay Cool

The simplest solution to this problem is to measure an equal distance from the top of each side of the milk carton and make a hole with a paper punch. Put the dowel through the hole and attach a windmill-like blade that has been cut from the plastic. On the other end, attach a handle that can be made from wood scraps and has a hole drilled in it to fit the dowel. Glue the handle onto the dowel. The important thing to remember is that the blade must be securely fastened so that it does not fly off when it turns.

♦ **Cross-Curricular Extensions** Look at the temperatures in different parts of the world to determine a good market for the product. Have students research how people adapt to or cope with the temperature in their environment. Do a class survey to find out how many students would prefer to live in a hot climate, a cold climate, and a climate with seasonal changes. Graph the results.

♦ **Materials** milk carton, dowel axle, large plastic lid or piece of stiff plastic, glue gun, wooden stick

Machines Design Brief 10
That's Not What It's For

This is a very open-ended design brief that is meant to challenge students' imaginations. Their final products will depend on the materials they have available to them. Before doing this activity, it would be helpful to have students replenish their materials with anything they can bring from home.

Some time taking apart old toys or machines would help them to become comfortable with seeing that everything is made from a combination of things put together. Safe use of tools should be reviewed before students start working on their inventions.

♦ **Cross-Curricular Extensions**   Have students look up the stories of some inventions and inventors from other cultures. Have them conduct a survey to find out about people's needs. Try to get a local inventor to visit the class. Look at science fiction stories to encourage futuristic thinking.

♦ **Materials**   coat hangers, egg beaters, spatulas, whisks, cutlery, clothespins, broken toys, pieces of plastic pipe, toothbrushes, combs, old appliances, clocks, jars, sandpaper

TOYS DESIGN BRIEFS

The toys design briefs can be used as the foundation for an integrated unit on toys or to complement a unit on play or manufacturing. You could set up your classroom as a toy factory. Toys would be an excellent unit to pursue during December when all the toys are being advertised or in June when students are thinking of things to do during summer vacation.

Toys Design Brief 1
Who's in the Box?

This design brief requires students to make a pop-up toy. Students actually have three tasks.

1. They must prepare a box with an attached lid that can be opened by the character inside. Boxes for China are perfect. A cracker box can be cut to the right size. Show students how to invert a box described in the Basic Techniques: Construction section on page 29. The box should be decorated so that it is attractive to the target age group. Students should remember that it is easier to decorate the box when they have it disassembled.

2. They must construct their character. They should have materials available such as toilet paper rolls, socks or mittens, paper cups, yarn, pipe cleaners, fabric, and colored paper.

3. Finally students will have to decide how they will get their character to pop out of the box. This can be done using pneumatics or springs. If they use springs, they must ensure that the lid of the box can be fastened securely. For a pneumatic solution, students should have had the opportunity to complete the pneumatics activities on page 35. Students should have access to the materials described there so that this power source is an option.

♦ **Cross-Curricular Extensions**   Have students compare the prices of toys in different stores and discuss how price is determined. Students should identify popular characters and discuss what makes them popular. Have students predict the results of student surveys and look at survey results for patterns. Ask, Are some patterns predictable?

♦ **Materials**   cardboard tubes, socks, mittens, paper cups, large and small syringes, plastic tubing, art supplies, glue, springs

Toys Design Brief 2
Play and Learn

This design brief would be an excellent assessment activity for a geometry unit on two-dimensional shapes and three-dimensional geometric solids. It could also be part of a physical education unit or a study of communities. The product will depend on

• the materials students collect

• their research on playgrounds and equipment

Once students have read the design brief and identified areas for investigation or research, they should be prepared to look more critically at a playground. A field trip to a nearby playground would be invaluable. You may also use architectural drawings or area maps.

♦ **Cross-Curricular Extensions** Discuss the careers involved in recreation design, planning, and maintenance. Examine the geometric features found in the school, playground, and park environments. Have students do measurement activities to demonstrate the size of playground equipment. Point out that architectural drawings are done to scale, usually $\frac{1}{4}$ inch = 1 foot. Focus on the safety features provided in playground equipment and recreation areas.

♦ **Materials** cardboard tubes, boxes, cardboard, string, adhesives, samples of two-dimensional shapes and three-dimensional solids, art supplies

♦ **Possible Solution**

Toys Design Brief 3
Designing Authors Who Help

A basic knowledge of bookmaking is required to complete this design brief. Students should also have access to children's books that have been put together in different ways, so that they can examine the methods of fabrication used. Students will then have to address the problem of making movable parts in their books. Students should examine the mechanisms used in movable children's books to see how to do pop ups or moving wheels with messages. Finally, in deciding upon the topic for their book, they must consider their goal. They want to show a child from another country what children here are interested in. Conducting a survey within the school will give students the information they need and will give them experience in gathering and managing data. The editing process used in any other writing assignment should be used here as well. As with any writing project, this brief will take several days to complete. It is ideal for integration with your regular language arts program.

♦ **Cross-Curricular Extensions** Find out what the completed shipment would weigh. Have students choose a destination. Look at options for transporting the shipment and determine the cost of shipping. Have students research the publishing industry and, in particular, how books are distributed.

♦ **Materials** bookmaking materials including construction paper, markers, fasteners, and other art supplies

Toys Design Brief 4
The Baby-Sitter

This is an unusual design brief in that it actually requires students to write a design brief for someone else. An effective way to introduce this brief is to do the example on the brief as a whole group activity. Have students build a safe wall for Humpty Dumpty. Afterwards, give each team a simple children's book. Give them about fifteen minutes to read the book and think of one activity that could be written to accompany that book. Encourage each group to share their idea with the class. You may wish to have them exchange books and think of a different activity for that same book. Students can usually come up with several ideas for the same book once they get started. After students have had an opportunity to explore this format, they should have no difficulty coming up with an activity for a book of their choice.

- ◆ **Cross-Curricular Extensions** Have students determine the cost of the materials the baby-sitter would need. Look at the rate paid to baby-sitters in the area, and discuss the possibility of getting paid for preparation. Examine the safety precautions that must be considered when preparing activities for young children. (For example, use non-toxic markers and blunt-edge scissors. Don't use peanut boxes or anything else that might cause an allergic reaction. Use nothing children might be tempted to eat.)

- ◆ **Materials** a variety of young children's books and design briefs for examination, egg cartons, bubble packing, Styrofoam chips, Lego®, cotton wool

- ◆ **Possible Solution** A sample solution might be a design for a house for the three little pigs, a toll booth for the three Billy Goats Gruff, a new chair or bed for baby bear.

Toys Design Brief 5
A Magical Toy

In making the thaumatrope, the main problem that students have to solve is placement of the pictures so that they merge when the card spins. The picture on one side must be upside down for this to happen. The exact location of the picture pieces on the card must also be considered. An effective method of doing this is to make a working model of their thaumatrope using see-through tracing paper or onion skin paper that they can then attach to the card with masking tape for testing. Once they are satisfied with their pictures, they can transfer them onto the card or plastic with a tracing wheel or by tracing over the pictures with dark permanent marker that will go through their paper. If the pictures are done on cards, they can be laminated for durability. The other criteria that must be met is that the toy must reflect the friend's three passions: animals, the environment, and magic. The toy they make can be called magical; the picture they choose could reflect animals in the environment. An animal in its natural habitat would be a good choice. Some students may decide to make their toy out of environmentally friendly materials, or they may view making the gift themselves as a way of addressing this.

- ◆ **Cross-Curricular Extensions** Students should develop spatial awareness by examining the placement of the two images that must be merged. Have students try designing a flip book.

- ◆ **Materials** rubber bands, cards or plastic lids for disks, art supplies

- ◆ **Possible Solution**

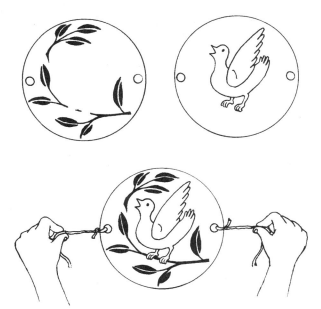

Toys Design Brief 6
Is This Math?

The first part of this design brief requires students to conduct a survey; they should come to this conclusion on their own. This will provide the information they need on the kinds of games that are popular. They then need an opportunity to examine a variety of games to decide if they can use an existing game in their solution. Once team members have decided on the kind of game they will make, the construction skills needed will have been acquired through regular art activities. They may need an adult to cut heavy cardboard with a utility knife; suggestions for using tools are given on pages 4–5.

◆ **Cross-Curricular Extensions** Reinforce measurement concepts by being more specific about the measurements of the game. Have students examine the probability of winning games. Ask, Can you increase your chances or is it luck? Have students make up a list of criteria for a comparison study of existing educational games and conduct the study.

◆ **Materials** math games and other games for examination, art supplies, utility knife, heavy cardboard

Toys Design Brief 7
Let It Roll!

It would be helpful if students had an opportunity to examine some toddler toys before they begin designing their own. The toddler's safety should be a major consideration for design teams when choosing materials. After considering the possibility that the child will take the toy apart, they should avoid having anything small inside the toy that a child could choke on. If they use bells or buttons for the noisemaker, they should string them together. One possible solution to this problem would be to remove the metal ends from a coffee can. (File or sand sharp edges.) Put some noise-making material inside. Cover the ends with plastic lids or cardboard. Make a hole in each end through which a handle could be inserted. Rubber washers could be glued on to keep the handle secure. The can could then be decorated with nontoxic paint or fabric. Another solution would be to leave one metal end on the can, and fill it with beans. Punch a hole into the can to avoid sharp edges and attach a string using a paper fastener. Attach the other end of the string in the same way to a plastic lid. Glue the lid onto the can for safety. Decorate the can. Students may come up with more creative solutions. The Basic Techniques: Construction section on page 28 shows the basic techniques students will need to assemble their toy.

◆ **Cross-Curricular Extensions** Have students design a print ad or commercial to sell their product. Discuss what price to charge if the product were sold. Have students look at comparable toys to compare prices and safety features.

◆ **Materials** cans, plastic lids, heavy dowels or long cardboard tubes, something to make noise, string, yarn, or laces, art supplies

◆ **Possible Solution**

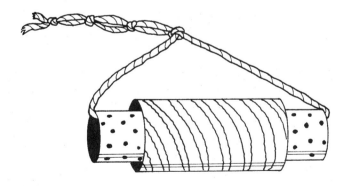

Toys Design Brief 8
What Happens When You Pull This?

It would be helpful if students had the opportunity to examine some infant toys, especially crib toys, before designing their own. Students are expected to design and make some kind of cause-and-effect toy. The safety of the baby should be the major consideration when team members are choosing their materials. They should be non-toxic and large enough so that they cannot be swallowed, and they should not be sharp. A simple solution would be to make holes in a large board and attach sets of matching objects to it. When you pull one of the objects, its attached counterpart should move.

♦ **Cross-Curricular Extensions** Reinforce students' understanding of symmetry or shapes by placing further parameters on the design brief. Have students make print ads or commercials for their product. Have them write a description of the advantages of their toy for parents.

♦ **Materials** wood or cardboard, plastic curlers, wheels, fabric, cardboard tubes, string, ribbon, art supplies

♦ **Possible Solution**

Toys Design Brief 9
Everyone Can Play

The results of this design brief will depend on the physical challenge the team decides to address, whether they choose to invent or modify a game, and which game they choose to modify. They should begin by researching the physical challenge they are trying to address. This could involve contacting specialists in the field or interviewing a physically challenged person. They should also have opportunities to role play by trying to play different games with the challenge they are addressing. Some examples of modifications they could make are

- using channels as guides on board games
- braille coding on cards, board games, or board game pieces
- providing taped directions for the visually challenged
- providing written directions or visual clues for those with hearing challenges

♦ **Cross-Curricular Extensions** Discuss the safety precautions that must be taken. Look at how the community provides for physical impairments. Have students write to local government officials for more information. Extend the discussion to include senior citizens. Ask, What are their special needs? Can students help?

♦ **Materials** board games, art supplies

♦ **Possible Solution** Students could make a bingo board with braille numbers, or they could make a board game with metal on the board and magnets on the game pieces.

Toys Design Brief 10
Please Come Out

You may wish to begin this design brief by brainstorming a list of animals and their homes. The brief could then be assigned in two parts with students completing their pneumatically powered animal first. Almost any animal can be made from toilet paper rolls. However, if you are making animals from some other material during art class, this would be an ideal way to integrate your technology. Once students have completed their animals, the Movement section on page 29 will provide the information students need to be taught in order to power their animals. You could then assign the second part of the design brief and have students use available materials to make their animal homes. Shoe boxes are ideal for dog houses, mouse holes, barns, bird houses, and so on. For wild animals, students could draw a habitat scene on their boxes.

♦ **Cross-Curricular Extensions** Have students write stories to go with their toy and then write directions for using the toy. Have students test their product on another class and analyze the market research feedback.

♦ **Materials** small boxes such as shoe boxes, plastic tubing, large and small syringes, art supplies

STRUCTURES DESIGN BRIEFS

The structures briefs can be used as the basis for a construction unit or as part of a unit that looks at communities, different cultures, or measurement.

Structures Design Brief 1
Structures All Around Us

In this design brief, students have an opportunity to explore how the world is made up of structures. They need the following background information.

- A structure is anything that can support its own weight and any reasonable weight that is placed on it, for example, people, objects such as chairs and tables, vehicles, towers, houses, bridges, trees, animals.

- There are two types of structures. A frame structure is made by joining materials together or framing the structure, for example, an electrical tower, a bridge, or a deck chair. A shell structure is held together by the material from which it is made: for example, a box, a can, or a tube.

- Many structures are a combination of the two. They have a frame structure underneath, which is then covered by a shell to form a very strong structure, for example, a house, our bodies, or a living room chair.

- Because structures must support their own weight and be freestanding, it is very important that they be designed so that their weight or the weight that is placed upon them is evenly distributed. The structure must have a strong base or be stabilized by using cross supports. A triangle is a very strong shape. Triangles are often used to provide stability in a structure. Cross supports often form triangles to strengthen a structure.

Once students have been exposed to these basic ideas about structures, they need an opportunity to explore the concepts. All of these design briefs provide this opportunity.

♦ **Cross-Curricular Extensions** Assign students different cultures to explore, having them take note of how different cultures use technology to meet their needs. Have students teach a younger class about their animal or insect and how its environment meets its needs. Discuss things we do that protect or harm environments.

♦ **Materials** construction paper, clay, pipe cleaners, string, yarn, wood scraps, boxes, bottle caps, Styrofoam packing materials, fabric scraps

Structures Design Brief 2
Geometric Structures I

This design brief introduces students to the concept of stability and the stabilizing effects of the triangle. Students then build a simple structure that they will enhance in the next brief. They should have a knowledge of basic construction techniques. They also need to be familiar with the names and characteristics of geometric solids and the difference between a frame and shell structure.

♦ **Cross-Curricular Extensions** Focus on measurement by putting dimension parameters on the problem. Look at how architects around the world have used geometric shapes in structures today and in the past. Look at the probability of groups coming up with the same structure.

♦ **Materials** wooden strips, corners, wood glue, bench hooks, saws, rulers, cardboard, paper fasteners, a set of three-dimensional geometric solids

Structures Design Brief 3
Geometric Structures II

Students have an opportunity here to see how geometric structures are being used in making other products. They can be very creative in deciding upon their product and are limited only by their imaginations and the materials available. No prior teaching is required for this brief.

♦ **Cross-Curricular Extensions** Prepare commercials to market the product. Videotape the presentations. Discuss the careers involved in marketing a product.

♦ **Materials** construction paper, fabric, light cardboard, yarn, buttons, pipe cleaners, stir sticks, art supplies

♦ **Possible Solution**

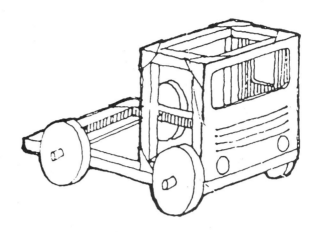

Structures Design Brief 4
Special Delivery

Students should have an opportunity to view or examine a variety of bridges, including arches, beams, and suspension bridges. They can find examples of such bridges in books and videos. Students should discuss what gives bridges their strength and stability. Students may construct their bridges out of straws, wood or plastic stir sticks, cardboard, rolled up newspaper, wooden strips, or commercial building sets. You may decide to limit their choices, depending on the focus you wish to take. To strengthen the structure, they may wish to use corrugated cardboard or paper rolls or accordion folded paper sandwiched between two layers of cardboard. If wooden strips are an option, the students should be familiar with basic construction techniques. Background information and introductory activities are provided on page 28. Students must think about how to strengthen their structure, especially if they are using a weak material like paper. They should investigate which shapes are strong, such as an arch or triangle. They must also design a reliable way to test the strength and stability of their bridge, such as using increasingly heavier loads and placing the loads at various points on their bridge.

- **Cross-Curricular Extensions** Have students examine the different architectural styles used at different times by different cultures. Discuss how the environment and natural resources affect the choice of materials. Look at the cost of building and maintaining bridges, why there are tolls, and jobs associated with bridges. Discuss how bridges have brought communities together. Look at the symmetry in bridges.

- **Materials** newspaper, wood and plastic stir sticks, cardboard, wooden strips, cardboard corners, straws, glue, tape, a toy car for each group

Structures Design Brief 5
Look Up! Look Way Up!

Students must spend lots of time on their design to be successful and efficient in this design brief. They should have the opportunity to tour an office building or the school to determine what must be included in their interior drawings. Try to get a local developer, contractor, engineer, or architect to talk to the class and show your students what a floor plan looks like. A woman speaker might encourage the girls in your class to consider these careers. Completion of a building's exterior structure requires a great deal of measuring as well as a working knowledge of area and perimeter. Much of the structure can be built from inverted boxes. Background information on this technique is provided on page 29.

- **Cross-Curricular Extensions** Have students make furniture and decorate the rooms. Put the class projects together to form a business park. Encourage students to investigate the different types of businesses that have offices in your community and the variety of available employment. Students can compare the cost of building a single-story house and a two-story house with the same square footage. Discuss the cost of real estate and land in your local area. Have students prepare a press release

announcing completion of the project and the anticipated benefits for the community.

- **Materials** cardboard, boxes, acetate, string, glue, tape, paint, wood scraps, commercial building sets if available

Structures Design Brief 6
Working Structures I

The solution to this design brief is some kind of stepping stool or box. However, students may think of some other structure that will solve the problem. The solution must be a structure. Students will need to have a knowledge of the basic techniques for construction. They will need to use cross beams to provide the necessary support and stability and must also consider the safety of the children in their design.

- **Cross-Curricular Extensions** Have students measure the children's stepping height. Have them graph and analyze the results. Students can design safety posters to go with the structure. Have them think of other devices that would be useful for people who need to reach high spaces that are otherwise inaccessible.

- **Materials** wooden strips, corners, saws, bench hooks, wood glue, safety glasses, paint, cardboard, construction paper, tape, art supplies

Structures Design Brief 7
Working Structures II

The emphasis in this design brief is on research. Students must find out about the many types of tower structures used for communication, fire lookout, observation, lighting, or other purposes. Classroom visitors who work in towers or construct towers would be a valuable resource. Students could generate a list of people or organizations that they could write to for information.

- ♦ **Cross-Curricular Extensions** Have students research information about towers. They can read fiction on a tower theme or put together a collage of tower pictures and interesting facts. Poetry can be written in tower form, that is, a one-word line followed by a two-word line, and so on. Students could make a graph to illustrate the height of famous towers or the towers made in class.

- ♦ **Materials** straws, pins

- ♦ **Possible Solutions** a lighthouse, a lifeguard tower, an electrical tower, a water tower

Structures Design Brief 8
Structure Gifts

If students do not recognize that bookends are the solution to this problem, you may want to have them look at how the same problem is solved in their classroom, in the library, and at home. The bookends are simple to make since the wood is pre-cut. Students only need to sand the wood, glue it, and tack it together firmly with nails. They should then test their products in case weight needs to be added. Stones could be added for weight and even made part of the decoration. Painting and decorating the bookends to suit the person who will receive them allows students to be creative. They could look at any available bookends to get ideas. The final task will be adding felt to the bottoms.

- ♦ **Cross-Curricular Extensions** Have each group become an assembly line to complete the first stages of the project before individual team members each finish a set. Have students calculate the cost of making bookends compared to buying them. Look at books and bookends from other cultures and discuss reasons for similarities and differences.

- ♦ **Materials** 5-inch pieces of 1-foot by 4-foot spruce, fir, or cedar, wood glue, finishing nails, varnish, paint, paint brushes, sandpaper, felt, stones, wood strip scraps

Structures Design Brief 9
Interesting Climbing

Students should have an opportunity to look at a variety of play structures either at playgrounds or in catalogues. They should also have some experience or some lessons on putting together and conducting useful surveys. Their climbing structure should reflect the survey results. It should contain as many features as possible to address the widest possible market. Their ad should focus on what makes their product special or different. Ask, Why would people want to buy it?

- ♦ **Cross-Curricular Extensions** Have students prepare an evaluation survey for students who will be watching their presentations. Have them graph the results of their original survey and their evaluation survey. Have them analyze the results of both surveys. Discuss the careers that involve promoting and maintaining fitness.

- ♦ **Materials** paper rolls, cardboard, string, plastic, straws, clay, stir sticks, newspaper

Structures Design Brief 10
Please Be Seated

Students can really use their imaginations on this product! Home decorating magazines and catalogues will help to inspire their creativity. Looking closely at chairs at school and at home will help students see what gives chairs their strength and stability. Distinguishing between which features give chairs their stability and strength and which features are purely aesthetic is an important part of this design brief. The number, shape, and placement of legs, use of cross pieces, choice of material, height, use of castors, swivels, joining techniques, and so on should all be noticed and discussed.

- ♦ **Cross-Curricular Extensions** Discuss what makes some chairs cost much more than others. Ask, What is the probability of two students in the class having the same kind of chair in their home? at school? Have students graph a survey of their favorite types of chair. Look at chairs from different cultures. Ask, How are they the same? Look at fabrics and their patterns. Discuss the manufacturing business.

♦ **Materials** cardboard, boxes, plastic, food trays, fabric scraps, wrapping paper, stuffing material, needles, and thread

ENERGY DESIGN BRIEFS

The energy design briefs provide opportunities for students to explore the sources, forms, and practical applications of energy.

Energy Design Brief 1
Tick Tock

You may wish to have students begin their investigations with pendulums by suspending the string from a desk and then experimenting with different lengths of string and different weights. Using a stopwatch, they can determine how long it takes their pendulum to swing back and forth. This is called the period. They can then try varying the length of the string while keeping the weight constant or varying the weight while keeping the length of string constant. They should record their findings on a chart like this.

String Length	Weight	Length of Period		
		Trial 1	Trial 2	Trial 3
6 inches	1 weight			
9 inches	1 weight			
12 inches	1 weight			
15 inches	1 weight			

String Length	Weight	Length of Period		
		Trial 1	Trial 2	Trial 3
12 inches	1 weight			
12 inches	2 weights			
12 inches	3 weights			
12 inches	4 weights			

They should see that the weight does not affect the period, but length of string does. To examine the height of the upward swing, they could stand a piece of graph paper behind the pendulum and try starting the pendulum at different positions. They should mark where it was released and where the swing ended. They should discover that the higher the starting point, the higher the swing. To heighten the upward swing, they must raise the starting point. Students are then ready to build a frame for their pendulum. They should be familiar with basic construction techniques that are described in the Basic Techniques section on page 28.

♦ **Cross-Curricular Extensions** Have students research the history of pendulums and how and where they are used today. Have them use their pendulums to time things.

♦ **Materials** string, different sizes of metal washers, wooden strips, corners, saws, bench hooks, eye hooks

Energy Design Brief 2
Teaching Others

In powering the gondola, students can trace the source of energy all the way back to the sun.

- The sun makes food grow.

- Food powers the lungs.

- Energy is transferred from the lungs to the balloon.

- The air is compressed in the balloon as potential energy.

- As the compressed air expands, the energy moves the balloon along the thread.

By experimenting with the angles of the attached string, students should discover that the kinetic energy of the balloon changes into potential energy sooner at steeper angles. They must also experiment with how much air to put into the balloon to enable it to travel the required height and distance. Adding weight to the gondola will cause the kinetic energy to be converted to potential energy much sooner. More energy is needed to move the gondola along the string. You may wish to have students make a balloon-powered car as an introduction to this activity. More details on this form of power can be found in the Basic Techniques: Power section on page 34. To build their gondola, students should use very light material, such as paper, folded crisply to form an open box. A straw can be glued to the bottom of the box. By threading the string through the straw, students provide a track for their gondola. The balloon can be taped to the bottom of the straw. The straw will travel along the string, powered by the balloon.

◆ **Cross-Curricular Extensions**   Have students research how and where gondolas are used. Have them decorate the gondolas.

◆ **Materials**   1 balloon, 1 straw, string, tape, paper or lightweight boxes, weights, art supplies

Energy Design Brief 3
Let It Blow

Students may have to conduct research on how a windmill works. They will need an electric fan to test their windmill. The wind (fan) will turn the sails or propellers. These are joined to the motor by the rubber band that goes around the spool and the spindle of the motor. The motor is attached to the light by the wires. Students need some knowledge of electricity and how to make a simple circuit. Introductory activities on electricity can be found on pages 36–38.

The materials you provide will give students some idea of what their windmill could look like. The cardboard blades should be joined with a round-head fastener like a propeller and glued in place so they remain spread out. Once this is done, students can bend the blades at the fastener to make them look more like blades. The pieces of card with the holes are glued over the holes in the milk carton for strength. The dowel goes through the holes in the milk carton and is secured in the back by a piece of tubing. In the front, the spool goes over the dowel and the propeller is glued to the spool and the tip of the dowel.

◆ **Cross-Curricular Extensions**   Have students come up with a name for their product. They should determine the cost of producing it and develop an ad to market it. Discuss the optimum market for this product.

◆ **Materials**   large juice or milk carton $\frac{3}{4}$ full with sand and a hole made through it (using a knitting needle), 5-mm dowel, $1\frac{1}{2}$-inch pieces of tubing, spool, 3-inch-by-2-inch pieces of heavy paper with a 5-mm hole punched in the center, 1.5V bulb, bulb holder, motor, wire, card strips for blades with holes punched in one end, round-head paper fastener, thick rubber band

Energy Design Brief 4
Let It Shine

Construction of this product is relatively simple. Testing and revising the basic design is the real challenge. Students must discover that some materials are better insulators than others. They should do the necessary research to find out the basics. Painting the can with flat black paint will increase absorption and retention of the sun's heat. You may wish to provide opportunities for students to experiment with these two concepts.

◆ **Cross-Curricular Extensions**   Have students investigate how solar energy is used in different parts of the world. Ask, Which areas of the world would best be able to utilize the sun's energy? Have students make a graph of the temperature recordings.

♦ **Materials** small and large cans, plastic food wrap, insulation materials (such as sand, Styrofoam chips, popcorn, shredded paper, and house insulation), thermometers, paint, play dough, hammer, nails

♦ **Possible Solution**

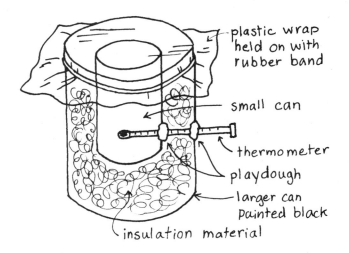

plastic wrap held on with rubber band

small can

thermometer

playdough

larger can painted black

insulation material

Energy Design Brief 5
Fun and Games

Students have an opportunity to explore hidden circuits in this design brief. They will need some background experience with electricity before attempting this product. Introductory activities for electricity are provided in the Basic Techniques: Power section on page 36. Students should also be concerned with the aesthetic value of their game and the language they use to word their questions and answers.

♦ **Cross-Curricular Extensions** Have students write explanations of how to use the hidden-circuit board. Ask students to make their

games to review ideas they are learning in history, science, math, or another subject. Have them organize a game day so that students from other classes or parents can try out their games.

♦ **Materials** cardboard, wires, buzzers, bulbs, bulb holders, tape, glue, paper fasteners, art supplies, batteries

♦ **Possible Solutions** A hidden-circuit board, like the one illustrated on page 38. It can have questions and answers right on the board, or there may be a separate page with questions and answers. Or students can make a board of true or false questions. True answers are wired to turn on a bulb when connected to the true button.

Energy Design Brief 6
Pick Me Up

Students will have to do some research to find out how to construct an electromagnet. They should discover that the strength of the electromagnet depends on the material they choose for their core, the voltage of the battery they use, and the number of times they wrap the wire around the core. After they construct their electromagnets, they will notice that the wires get hot when it is being used. This is caused by the friction created when the electrons move through the wire and rub against the atoms in the wire.

♦ **Cross-Curricular Extensions** Have students conduct research on where electromagnets are used (car starters, cranes in junk yards, doorbells, telephones). Have them make a presentation to another class on how electromagnets work.

♦ **Materials** batteries, insulated wire, materials to try using as the core (nails, straws, wood)

♦ **Possible Solution**

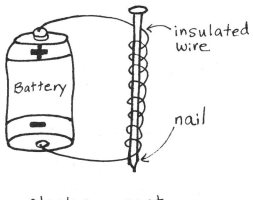

electromagnet

Energy Design Brief 7
Hover Power

This vehicle is simple to construct, but students will have to experiment with a variety of disks and balloons to discover which combination produces the desired results. Students should decorate their spool and disk and choose the color and shape that they feel is most attractive for the balloon.

♦ **Cross-Curricular Extensions**   Have students graph the results of their trials. Have them prepare an ad for their vehicle. Have them discuss the real-life applications of such a vehicle.

♦ **Materials**   balloons, thread spools, Styrofoam trays, lids, paper and plastic plates, clothespins, art supplies

♦ **Possible Solution**   The spool can be glued to the plate, tray, or lid. The balloon goes through the spool and plate.

Energy Design Brief 8
Tourist Attraction

Students need to know how to make a wheel and attach it to an axle. This is addressed in the Basic Techniques: Movement section on page 29. The water wheel must be mounted on some type of frame that could be of simple wood strip construction. This is addressed in the Basic Techniques: Construction section on page 28. The advertising could be done on an attached card disk.

♦ **Cross-Curricular Extensions**   Have students make a brochure for their water park. Have them make a list of all the jobs that would be involved in such an operation.

♦ **Materials**   large lid covers, bottle tops or pieces from a plastic bottle, cups, nails, glue gun, wooden dowel, rubber washers or plastic beads

Energy Design Brief 9
Gravity Power

Students should have learned basic construction and movement techniques in preparation for this design brief. These are addressed beginning on page 28. Their triangular tower should have a spool over which to run the thread that contains the weights. The thread should be wound around to the car's axle. Students should be able to wrap the thread around the axle a few times and then wind the wheels until the weight rises to the spool. When they release the wheel, the weight should cause the thread to unwind and push the vehicle in the opposite direction to which the thread has been wound. Students should experiment with the length of the thread and the amount of weight they wish to use.

♦ **Cross-Curricular Extensions**   Have students time their runs and chart the results of different trials. Have them look at vehicles from different cultures and discuss how they are designed to meet specific needs.

♦ **Materials**   wooden strips, corners, bench hook, saw, drill, glue, washers, thread, dowel or lollipop sticks, wheels, thread spools

♦ **Possible Solution**

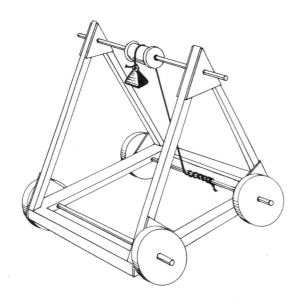

Energy Design Brief 10
Now We're Cooking

This is a simple design brief that requires very few materials. It provides students with an opportunity to observe the power of the sun's energy. Students can be as creative as they wish in their designs. Some may even wish to power the moving parts of their cookers, using electricity. They must consider these safety features.

- Do not look at the sun or its reflection.

- Wear protective clothing to prevent burns.

- Wear sunscreen to avoid sunburn.

♦ **Cross-Curricular Extensions**   Have students design safety posters. Discuss how technology has resulted in damage to the ozone layer and what this means for future generations.

♦ **Materials**   boxes, aluminum foil, mirrors, coat hangers, metal skewers, marshmallows

TRANSPORTATION DESIGN BRIEFS

The transportation design briefs are based on the world of work. It is important that students make connections between what they learn and applications for the real world. They must be given opportunities to reflect on the kinds of skills and attitudes required to become successful, contributing members of society. In these design briefs, students will be required to design a variety of vehicles, but also to set up companies and have their vehicles accomplish specific tasks.

Transportation Design Brief 1
Message from Above

In this brief, students explore the effects of gravity and air resistance. As a parachute falls, the earth's gravity is pulling it down, but the air resists its fall. The earth's gravity is a much greater force than the air resistance. As students design their parachutes, the objective is to create as much resistance as possible. The more resistance students can get their parachutes to create, the slower it will fall. In commercial parachutes, nylon fabric is used to make the canopy. It is usually round or rectangular and has a large enough surface area to provide the correct amount of air resistance. The weight of what the parachute is carrying also affects the speed with which it falls. As students make their parachutes, they should be encouraged to experiment with different canopy sizes and shapes, different cargo weights, different lengths of string to attach the canopy to the cargo, and different wind conditions. It is also important to have a hole in the center of the canopy so that the air will pass through.

♦ **Cross-Curricular Extensions**   Have students write the note the grounded student might have written. Have them time the fall of their parachute during several tests and record the data in a graph.

♦ **Materials** string, plastic bags, lightweight fabric, film containers, tissue boxes

Transportation Design Brief 2
Ready! Aim! Fire!

The catapult is a simple mechanism to construct with the materials provided. Some students may have to research what a catapult is or what it looks like. They will have to be familiar with the basic techniques for construction on page 28. The real challenge for students lies in calibrating a piece of cardboard or designing some other type of measuring device that will allow them to regulate the distance the load travels. The distance traveled depends on how far back they pull the arm before releasing it. By conducting several test fires, students should be able to determine with some degree of accuracy the exact distance for each position of the firearm. To help with classroom management, you may want to designate a test area.

♦ **Cross-Curricular Extensions** Have students turn their catapult into a game with a target. Do an investigation into exactly how catapults were used in medieval times. Have students stage a medieval battle, using toy figurines and building their own set.

♦ **Materials** wooden strips, skewers or dowel, lid covers or plastic cup, rubber bands of different sizes, clothespins, tacks, marshmallows, 8-inch pieces of 2-inch by 4-inch wood, screw-in hooks, other gatherable materials.

♦ **Possible Solution**

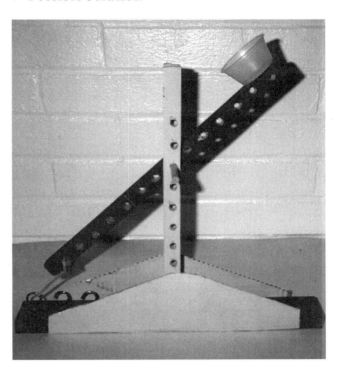

Transportation Design Brief 3
Fly with Us

Students should have opportunities to view a variety of print as well as TV ads before attempting their own. Encourage them to notice the sound effects, the emotional appeal, the choice of actors or characters, the information given, and the target audience. They will need practice in writing scripts that convey a message in a short time segment. Acting out the commercial in a convincing way will require many rehearsals. Access to a camera is important so that students can critique their efforts in advance.

♦ **Cross-Curricular Extensions** Have students perform their commercials for parents or another class. Have them design a logo for their airline. Have them find out the cost of running a one-minute ad at different times of the day.

♦ **Materials** TV commercials, video camera

Transportation Design Brief 4
Space Robot

Students can be very creative with this design brief. The robot design will vary, depending on the task the robot will perform. Required prerequisite skills will also vary. Students may choose to have some part of their robot consist of a simple machine that will perform a task. Using wheels or pneumatics would be another means of giving the robot its capabilities. Students should research the tasks robots are presently performing but they should not be limited by this information. Even fictional robots would provide a source for ideas to get students started.

♦ **Cross-Curricular Extensions** Have students write a Work Wanted ad for their robot. Discuss the implications of robots on the work force, including loss of jobs and less exposure to dangerous or monotonous tasks. Do a survey of the community to find out if and how robots are being used.

♦ **Materials** cardboard boxes, pipe cleaners, tape, paper clips, syringes, balloons, plastic tubing, rubber bands, glue

Transportation Design Brief 5
Wanted: Space Project Workers

Students should have access to as many examples of Help Wanted ads as possible in preparation for this design brief. They should analyze a variety of ads, focusing on the information provided in each ad, how it is worded, and what makes it different. Viewing futuristic videos or reading futuristic books would be invaluable to students as a source of ideas about the kinds of careers that will be available when they grow up. Their ad is being prepared for publishing and should be edited by the team for perfect copy. The criteria for quality expected for any work that is to be published should be expected for this piece of writing.

♦ **Cross-Curricular Extensions** Have students write to some experts for career forecasts. Organize a futuristic career day complete with costumes.

♦ **Materials** futuristic videos or books, a variety of Help Wanted ads

Transportation Design Brief 6
I Missed My Train

Once students have researched and made a list of everything they will need to include in their model, they will be ready to begin construction. A wide variety of boxes should be available for this design brief. You may wish to have students start collecting them from home a few weeks before. The teams may also want to bring in model trains and other vehicles, as well as small characters to make their models look more authentic. You could have the teams build everything, but it would take a long time. The focus in this brief is on examining the intricate workings of a large transportation system.

♦ **Cross-Curricular Extensions** Have students create rail tickets and passes. Make a chart of the occupations involved, how much education is involved for each, and the salary that can be expected for each job.

♦ **Materials** a variety of boxes, paper rolls, ice cream or stir sticks, art materials

Transportation Design Brief 7
Blast Off!

This design brief requires futuristic thinking, and students would benefit from viewing futuristic videos or action comics. This brief would combine well with Transportation Design Brief 5. The same preparation could be used for both.

♦ **Cross-Curricular Extensions** Have students investigate how present-day spaceships are powered and how people survive in space. Students can write to NASA for information on space travel. Discuss the implications of space travel for family life as we know it.

◆ **Materials**   boxes, paper rolls, pie plates, aluminum foil, cans, wire, pipe cleaners, buttons

◆ **Possible Solution**

Transportation Design Brief 8
Puff Power

The best material to use for this car is a milk carton with card wheels glued onto skewer axles. One wheel should be glued to the axle, the axle pushed through holes in each side of the carton and the other wheel glued on. The balloon can be inflated and the end pulled through a hole in the back of the carton and held closed with a clothespin, string, or paper clip. The size of the hole will determine how quickly the air is released. Background information on making wheels is provided on page 29.

◆ **Cross-Curricular Extensions**   Have students measure the distance traveled by the car. See if it can climb an inclined plane. Have them try to control the distance traveled by counting the number of puffs of air they put in the balloon.

◆ **Materials**   milk cartons, wooden skewers, heavy cardboard for wheels, balloons, paper clips, clothespins, wooden strips, lids, corners, bench hook, saw, glue

Transportation Design Brief 9
Move It and Dump It

The first task in this design brief is to build a dump truck. It can be built from wooden strips and boxes or completely from boxes. The dump box will have to be hinged so that it can be raised and lowered, and it will need a hinged tailgate that can be opened and closed. Students will need to know the basic wood and box construction techniques on page 28. Once the truck is completed, they must install a pneumatic power system to operate the dump box. Introductory activities are provided on page 36. A deflated balloon attached to the syringe and placed under the box will allow it to raise and lower.

The team must also set up their company and make plans for completing the job. Since efficiency is a consideration, it will be interesting to see if the teams think of combining the trucks to make a fleet that could complete the job more quickly. If they look at the time involved and the cost of overtime, they may see advantages to combining or leasing trucks. If you can actually set up in detail the situation described, it would be an excellent scheduling activity. For example, recesses or certain periods may be the only times they have access to the site.

◆ **Cross-Curricular Extensions**   Discuss the following questions. What would be the cost of labor? How much sand is an optimum load? How much sand will the truck hold, carry, and dump? How long would it take to complete the job with one truck, two trucks, or more?

♦ **Materials**   wooden strips, corners, saw, bench hook, glue, boxes, syringe, tubing, balloon, wooden wheels, skewers, rubber washers or beads

♦ **Possible Solution**

Transportation Design Brief 10
Public Transit

Students may come up with original designs or traditional bus designs. The axle and wheels can be attached by punching holes in the corners and gluing them on the outer sides of the bus. Background information on making and attaching different types of wheels is described in the Basic Techniques: Movement section starting on page 29.

♦ **Cross-Curricular Extensions**   Students should look at the cost of traveling by public transit compared to individual cars. Have them examine the impact of each on the environment and make posters promoting the use of public transit. Discuss the issue of accessibility by everyone.

♦ **Materials**   shoe boxes, small boxes, cardboard for wheels, art supplies, skewers, corners

What's in a Name?

Throughout this unit, you are going to be approaching problems as if you were part of a company. Every company has its own style, values, and goals. Every company also has a logo, a way of identifying itself. Sometimes the logo simply displays the company name. It may also be designed to say or show what the company does. Sometimes the logo communicates the company's values. Successful companies have a vision of what they want people to think of when they see the company logo.

Suppose your company works as a consultant to other companies, helping them market their products. Design a name and a logo for your company. You should begin by choosing a vision for your company. Decide what its style, values, and goals will be. Your logo should help people remember your company.

Name That Product

One of the most important aspects of bringing a product to market is deciding on a name. The product name should be easy to remember, such as Big Mac®, or it should create an image in the consumer's mind, such as Raid®. The name might focus on what the product does or promises to do, as 2000 Flushes® does; it might emphasize some characteristic of the product, as Sugar Crisp® does; or it may be just a catchy word or phrase, such as Frisbee® or Dirt Devil®.

Create a name for these two products: a low-fat chocolate bar and a child's construction set. Draw pictures of each product.

Communication Design Brief 3
Print Ad Campaign

Once you have chosen a name for a product, you must decide how to let people know it is available. You also must persuade them to buy it rather than any of the competing products. Choose one of the products you named. Prepare a print ad campaign to promote the product. There are many ways to advertise. Billboards, posters, or ads in magazines or newspapers are popular print media. They are often less expensive than television, although they also reach a large number of consumers. Examine a variety of print ads to decide what you should include in your own ad.

Things to consider

- How did it catch your eye?
- What made you want to read it?
- What information did it include?

Don't Change That Channel

Radio and television are media that allow you to give the consumer a more active and creative message about your product. The characters you create, the voices, jingle, or slogan you use, and the scenario you design when you write the ad will all have an impact. With an active media, you can communicate with the consumer through both sight and sound. You can create a vision or a fantasy that might appeal to a consumer's emotions, and you can focus on the practicality and convenience of your product.

Prepare a radio or television commercial for one of the products you named in Design Brief 2. Videotape or record your commercial so that you can review it and make any necessary changes.

KidTech © Dale Seymour Publications®

Follow That Ad!

Moving ads are a creative way to catch the consumer's attention. Moving ads are often found on billboards and store signs or at airports and theaters. They capture your attention by revealing only part of the message at a time, so that you want to keep watching. Often you cannot tell what is being advertised until the very end of the message. Sometimes moving ads are used to advertise an upcoming event.

Design and make a moving ad for a product you named in Design Brief 2 or for an upcoming event in your school or community. Your ad can be one that must be moved by hand or one that moves mechanically. You may even wish to light your sign. Think about what color combinations would attract the most attention.

Market Analysis

When you are designing a product, you should always have a target market in mind. One way to decide on a target market is to look at products similar to yours that are already on the market. Then see if there is a gap in the market that you could fill.

Things to consider

- Which market range is already being targeted by other manufactures? Why do you think this is so?
- Which manufacturer has the best buy?
- How do top-of-the-line manufacturers justify charging what they do for their product?

Do a market analysis of the cars that are available in your area. Choose three or four different manufactures and find out the range of cars they offer. Organize your findings in a chart similar to this one.

Manufacturer		A	B	C	D
Price Range	Top of the line (over $40,000)				
	Above average ($30,000 – $40,000)				
	Average ($15,000 – $29,999)				
	Basic model (under $15,000)				

Look at your findings to see if there are any gaps in the market. What market would be a good one to target if you were building a car?

What Does It Cost?

When providing goods and services, you must always be conscious of your costs. The costs involved vary with the product or service, but in every case there are many things to consider. You must consider all your options carefully if you hope to make a profit. In this design brief, you have two choices.

1. Plan and offer a lunch at your school.

2. Plan a craft sale in which your class makes the crafts and sells them.

Keep track of your costs and how much you must charge to make a profit.

Things to consider

- workers' hourly wage (Use minimum wage.)
- supplies needed
- advertising costs
- space, tools, and equipment
- sales tax owed
- utility costs

What if you limit the choices you give customers, buy in bulk, or use an assembly line? How will these options affect your costs?

Consumer Survey

Many publishers publish children's books written by children. Your task in this design brief is to write a book for children five to seven years old. Your book should be unusual in the way you put it together or the finishing touches you add. To decide on the content for your book, survey your target audience to find out the following information about their preferences.

- the kinds of stories they like
- the kinds of characters they like
- the size book they like to handle
- what makes them notice a book or keep going back to it

   Remember that your book will be marketed and must compete successfully against many other books. The content of your book, the illustrations, and the cover must look professional.

Communication Design Brief 9
Promotional Flyer

Now that you have a book to sell, you must think about the best way to let your target market know it is available. Many publishers send out flyers to advertise a new book. The flyer tells something about the book, something about the author, and when the book will be available. It is designed to catch the attention of the market. Design and make a flyer to advertise your book. Where would you send your flyers? Where and when will your book be available? Are you planning a book launch where you will be available to sign copies of your book?

Does the End Justify the Means?

Companies that provide the goods and services needed by society must be concerned about more than making money. The development of technology has positive and negative effects on the planet. We have a responsibility to be informed about the results of our actions so that we can make wise decisions. Prepare a presentation that outlines the positive and negative effects of technology on our planet. You might include information about some of these things.

- the ozone layer
- wildlife
- the greenhouse effect
- automated farming
- fast foods
- use of robots
- natural resources

- pollution
- rain forests
- landfill sites
- use of energy
- organic farming
- developing countries

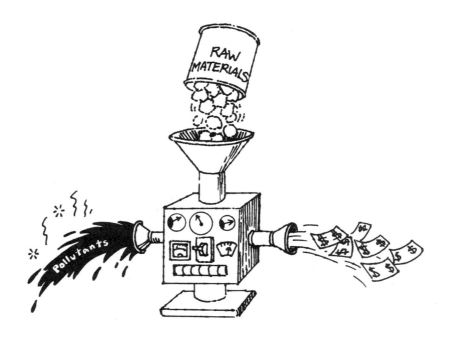

Machines Design Brief 1

Time Is Up

You are trying to help the kindergarten teacher who wants the five-year-olds to become more independent about moving from one activity center to another in the classroom. The teacher wants some kind of timing device to let students know when 15 minutes is up and it is time for them to move. These students cannot tell time! An egg timer might work, but wait! You have just learned about levers in science class, and you wonder if your knowledge about levers might be useful in this situation.

Design and make a see-saw timing device that uses your knowledge of levers and egg timers. Remember that your device is meant to help the five-year-olds become more independent, so they should be able to operate it on their own.

Save That Kitty

You are out hiking with your dad when you hear a whimpering noise from a ledge below you. When you look down, you see a small gray cat. Its leg seems to be broken, and the cat is stranded on the ledge. After a difficult climb, your dad reaches the cat. He knows that the cat will need to be taken to a vet, but first he must get him up the hill. He has no knapsack, and the cliff is too steep to climb with the cat in his arms. He can use his shirt!

Design and make a cloth carrier that you can operate from the top of the hill to help your dad. Your dad will have to stay on the ledge to secure the cat in the carrier. Try to raise a stuffed animal from the floor to your desk to test your system.

A Changing Sky

Your class runs a puppet theater for younger students to attend during recess on rainy days. You are responsible for designing the backdrops for the plays. The play you are working on now calls for a night sky with a moon and a day sky with a bright sun. There is no time to change the backdrop between scenes.

Design and make a backdrop that uses a belt and pulley system to make the sun and moon rise and set. Make your backdrop as attractive as possible; you want your stage set to look professional.

Things to consider

- Should the pulleys be the same size or different sizes?
- How should the sun and moon be placed so that when one goes up the other comes down?

Machines Design Brief 4
Open for Business

Your little brother has a lemonade stand. You would like to show off your technological skill by installing gear-operated shutters. Use a box with a cutout window as a model of the lemonade stand. Design and make a model gear system that your brother will be able to use to open and close his shutters. Make the shutters say "Closed" when they are shut, and have them display an ad for lemonade when they are open.

KidTech © Dale Seymour Publications®

Machines Design Brief 5
Talking All Night

Summer vacation is here at last, and your family is going camping. Everything is perfect, or at least you thought it was. On your first day at the campground, you make a great new friend. Your friend's family is camping in the next campsite. You have not stopped talking since you met! You have so much in common. Lights go out at the campground at ten-thirty. Everything is supposed to be quiet after that. You are always wide awake then and would love to talk to your friend. How can you communicate without disturbing other campers?

Design and make a pulley system that will allow you and your friend to send notes back and forth.

Your Baby Sister's Pool

Going to a cottage at the lake is always one of the highlights of your summer. Fishing, swimming, water skiing, and jet skiing keep you busy all day long. Your little sister is too small to go into the lake, so your parents have bought her a kiddie pool. The cottage has a well but no outdoor tap. You have been given the job of filling the pool from the lake. You're afraid it will take all day to do this. How can you make your job easier?

Design and make a water wheel that will allow you to send water from the lake down a chute into your sister's pool. Test your invention, using a large dish of water as the lake and a smaller dish to represent your sister's pool.

Machines Design Brief 7
Next Please

Imagine if your class had to make 5,000 mini pizzas with exactly the same ingredients on every one! The job would go something like this.

1. Take a plain pizza crust and put sauce on it.

2. Pass it to the next person, who puts on the cheese.

3. Pass it to the next person, who puts on pepperoni.

4. Pass it to the next person, who puts on mushrooms.

5. Pass it to the next person, who puts on green peppers.

6. Pass it to the next person, who puts it in the oven.

This is how work is done in factories. A conveyor belt saves time by moving the product from station to station. Design and make a conveyor belt system that could help accomplish a repetitive task.

Pick That Up

Think about how often you have to bend down to pick up something. Nobody likes doing it, and it gets more difficult as you get older. We have to pick up things we drop; we pick up litter; we pick up after our pets; we pick up leaves; and of course we pick up special things we find on the beach.

Design and make a machine that will help people pick up a variety of things without having to bend down. When you have completed your product, prepare a 1- to 2-minute commercial to sell it. Remember to give your product a suitable name and a competitive price.

Stay Cool

Do you ever find it really hot in your classroom, at home, on the bus, or in a car, only to find out that everyone else is comfortable or even cold? What you need is a personal cooling device that you can carry with you anywhere!

Design and make a personal cooling device that you can operate by turning a handle. Think of an appropriate name for your device and prepare a newspaper ad or a billboard to market your product.

Things to consider

- Is it important that the cooling device be attractive?
- Whom do you want to target in your advertising?

That's Not What It's For

Many inventions happen by accident. Often an inventor begins with something we already have and finds a new way to use it or change it in some way to make it better. This is what you are going to do in this design brief. In order to do this, you have to open your mind to new possibilities. Look at some of the objects your teacher has given you and try to forget what they are used for normally. Combine two or more objects to make something new. Give your new product a name and be prepared to describe how it works and what it is used for.

Things to consider

- What are the characteristics of your original objects?
- How might those characteristics make the new object useful?

Who's in the Box?

Toy manufacturers must anticipate how children's interests and tastes change. You have decided to go into the toy business. You want to improve on a toy that you know has always been popular. You have chosen the jack-in-the-box. To make it more appealing, you will replace Jack with one of today's popular characters.

Design and make your own pop-up toy.

Things to consider

- What age group will be using the toy?
- What characters are popular with that group?
- How will you make the box attractive?
- How will you make your character?
- How will you make your character come out of the box?

Toys Design Brief 2

Play and Learn

The recreation department in your community is building a new playground. They have asked people to submit ideas for the playground design. You know you will be spending a lot of time there with your friends, so you really want to get involved. You have given it a lot of thought, and you want it to be more than a place to play. It could be a great place to learn geometry! You have decided to submit a plan. Design a geometric playground. When you are satisfied with your design, try to build it. Use whatever materials are available inside or outside your classroom. Include as many two-dimensional shapes and three-dimensional solids as possible.

Things to consider

- What kinds of equipment do you want in your playground?
- What does a playground need besides equipment?
- What natural features, such as landscaping, will it have?
- What will be the layout?

Toys Design Brief 3
Designing Authors Who Help

Your school has adopted a school in a developing country. You are putting together a care package. Every class is doing something special to contribute. You know that the school has very few books, and your class has decided to contribute books. The design teams in your class are each making a book to contribute. You want to make your book very special. You want the book to show a child in another country something about the children in your country and what they like.

Design and make a book that another child would enjoy. Some of your characters or scenes must have moving parts or hidden text.

Things to consider

- What age group will you write for?
- How will you make your scenes or characters move?
- What kind of text should be hidden?
- How will you put your book together?
- How will you make sure that your book lasts a long time?

The Baby-Sitter

You have been asked to baby-sit a six-year-old. You know that keeping him entertained is going to be a challenge. You also know that the child is just learning to read and loves nursery rhymes and fairy tales. The mother would also like you to involve the child in some kind of creative activity.

Choose a fairy tale or nursery rhyme to read and then design an activity that could be used as a follow-up to the rhyme or story. For example, design a way to keep Humpty Dumpty from hurting himself. Write a detailed description of the activity and then test your activity.

A Magical Toy

You have been invited to a birthday party. Guests have been told that all gifts must be homemade. The party is for a friend who has a passion for three things: animals, the environment, and magic! You want to design a gift that your friend would love. You decide to make a thaumatrope, which is an optical toy. It is made by putting parts of a picture on both sides of a hard, flat piece of cardboard or plastic. There is a hole on each side near the edge to which rubber bands have been attached. When you wind up the rubber bands, the object spins and the two parts of the picture merge as if by magic!

Things to consider

- Your thaumatrope should be attractive.
- It should reflect your friend's three passions.
- It should be durable to last a long time.
- How should you place the pictures?
- What do you want your merged picture to be?
- What parts of the picture will you put on each side?

Is This Math?

Math can be a difficult subject to teach well. Teachers and educational toy manufacturers are always trying to think of new ways to help students with this subject. Design and make a game that could be used by students to practice some mathematical skill or concept. You may use any materials available. Your game should be attractive and should include a container that will hold the game pieces. Remember to write instructions for your game and give it a name.

Things to consider

- What kinds of games are popular with students?
- What math skill or concept would you like to address?
- Can you modify an existing game to meet your needs?
- Can you use an existing game for a model?
- Can one of your computer games be used as part of your solution?

Let It Roll!

People start experimenting with sound as soon as they are born. Babies entertain themselves by listening to their own gurgling noises and are fascinated by musical toys and people's voices.

Design a toddler's toy that will make noise when it rolls. The toy should also encourage the child to crawl or walk. Consider using a can with plastic lids or heavy long cardboard tubes with something to make noise. Use string, yarn, or laces, and whatever art supplies you have to decorate the toy. Think about what safety factors you should consider when making a toy for a small child. Remember to give your toy a name.

Toys Design Brief 8
What Happens When You Pull This?

Babies spend much of their first year in a crib. Parents are always looking for toys that will keep their child happy in the crib. They also look for toys that will help in the child's development.

Design an educational toy that could be attached to the side of a baby's crib. The toy should do these things.

- reflect concern for the baby's safety
- encourage the baby to explore cause and effect; for example, pulling something could cause something to happen
- allow the baby to control the cause and effect
- introduce the baby to some basic concept, such as color and shape
- encourage the development of the baby's grasping and pulling skills
- attract the baby's attention

T o y s   D e s i g n   B r i e f   9
Everyone Can Play

Many of our favorite board and card games cannot be used by children who are physically challenged. We can do something about that! Adapt a game that you have in your classroom or at home, or invent a new game that could be enjoyed by a physically challenged student.

Things to consider

- What physical challenge are you addressing?
- Are you doing this for someone you know?
- Are some games easier to adapt than others?
- Do you have to change the rules?

Please Come Out

Toy inventors get most of their ideas from real-life situations. Dolls, toy vehicles, building sets, playhouses, baking sets, and cuddly animals are only a few examples of toys that are based on real life. The more the toy resembles its real-life partner, the more popular it seems to be. Your task in this brief is to design and make a toy animal and its home. You must use pneumatic power to enable the animal to come in and out of its home. You may use any materials available. Remember that your toy is competing with millions of others for its share of the market. Make it attractive to children and give it a name that will help it sell.

Structures All Around Us

We are structures, and we live in a world of structures. A structure is both

- freestanding
- able to support its own weight and any reasonable weight that might be
  put on it

Look at any animal or insect in its environment and count the many structures you see. Choose any animal or insect and build a model of it in its environment. Include as many structures as possible. When you are choosing your animal or insect, do not limit yourself to the ones you are familiar with. Think about the whole world, think about variety, and then choose!

Geometric Structures I

Many structures have geometric shapes. These shapes can help give a structure strength and stability. Conduct this experiment.

1. Make a square or rectangle, using four strips of cardboard and four paper fasteners.

2. Now squeeze the sides in so that the shape becomes a diamond.

3. Next, reshape it into the square or rectangle.

4. Add a diagonal strip of cardboard to your original shape, so that two triangles are formed. Attach it with the corner paper fasteners.

5. Now try to squeeze it into a diamond. You can't! The final strip you added was a crosspiece, and it made the shape stronger and more stable.

   Now build your own geometric frame structure. It can be a cube, a prism, or a pyramid.

KidTech © Dale Seymour Publications®

Geometric Structures II

Now that you have constructed a geometric solid structure, look for places where this structure is being used. Is there anything in your classroom that has this same frame?

Use your frame structure to create something interesting. It can be a character or an object; it can remain a frame structure, or you can cover it with a shell. When you are finished, write a description of your creation and give it an appropriate name. Be prepared to discuss how this product is going to affect society.

Special Delivery

Schools were not always as relaxed an environment as they are today. There was a time when students were allowed to leave their desk only to sharpen their pencil, go to the bathroom, or go to the teacher's desk! Students were not permitted to go to a classmate's desk, not even to borrow an eraser! Imagine yourself in a classroom like this, except for one thing—you have a teacher who thinks it is important that you share. The teacher tells you that you must follow the school rules, but encourages you to figure out a way to share your things with the student who sits in the next aisle.

Design and build a bridge that will allow you to send pencils, erasers, scissors, and other small things to the student who sits across from you. You may use a toy car to transport things, but it must be able to travel on the bridge between your desks. The space between your desks is 25 inches. The weight your bridge must be able to support depends on the car you use and the weight of the materials you transport.

Look Up! Look Way Up!

When building a structure, especially in a heavily populated area, space is often a major consideration. This reality is also influenced by the cost of land. Engineers and architects may be faced with the challenge of providing 25,000 square feet of space on a single city lot of 3,000 square feet. The only way to do this is by building up!

Imagine you have been hired by a land developer in Mini City to build an office tower that provides 8 square feet of space. Your city building lot is 1.5 feet square, and Mini City requires that all buildings have at least 3 inches of landscaping on all sides. The developer also insists that the building be interesting to look at, have an underground parking lot with space for 30 cars, and provide space for 40 offices. Remember to consider what other kinds of spaces will be needed in the building. Design and construct a model of the exterior of the office tower and grounds. Do drawings of what the inside of the building will look like.

Working Structures I

Look at the structures around you and think about which ones are designed to be decorative and which ones are designed to be functional. Many structures have been designed as a solution to a problem. For example, a short lighthouse might not be very effective. Design and build a model of a structure that could be used by the kindergarten children to help them reach things on a high shelf. Remember that a structure must be able to support its own weight and the weight that will be placed upon it. Test your structure by using a 2-pound weight. Consider the fact that although this structure is being built to solve a problem, it is going to be used in a kindergarten classroom by kindergarten students. Think about how it should look.

Working Structures II

Towers are fascinating structures that have been used to solve problems for centuries. Make a list of all the towers you have seen or read about and what they were used for. Towers are a challenge to build because they are so high. Your task is to design and construct the highest tower you can, using only straws and straight pins. Your tower should have a purpose. In your product presentation, be prepared to describe what your tower will be used for and how it will be used to do this job. Also be prepared to describe what materials will be used to construct this tower, how the materials will be joined, and what effect this tower will have on the environment.

Structure Gifts

Structures make wonderful gifts! Make a list of ten structures that members of your team have received as gifts. (Remember the definition of a structure!) Design and make two matching structures that can be used to keep books from falling off a shelf. You will be giving them to someone as a gift, so it is important that you consider the tastes of the person you have chosen. Try to think of a way of using what you know about the person to make the gift personalized. Be sure to test your product.

Interesting Climbing

Everybody today realizes the importance of keeping fit and that fitness starts when you are young. Exercise is an important part of fitness, and people are more likely to exercise if it is fun. Playground designers and exercise equipment manufacturers take this into account when they are designing their products. They often survey their market to find out what people like.

Conduct a survey of a group of five- to eight-year-old students to find out what they enjoy in a climbing apparatus. Design a model of a climbing apparatus that will appeal to your target market. Construct your model, using materials you have gathered, and then prepare an ad to sell your product. Present your ad and your product to the survey group to find out how successful you have been at meeting their needs.

Please Be Seated

Some of the everyday things we take for granted are, in fact, amazing examples of very carefully engineered and constructed structures. The chair is a perfect example! A chair has to be designed so that it not only supports its own weight but the weight placed upon it. A chair also has to be attractively designed to compete in the furniture market. Design and construct your own model chair. Your model should be less than one foot tall, and it must be able to support a weight of 2 pounds. Your chair must also be attractive. Remember the competition!

Energy Design Brief 1
Tick Tock

Energy causes things to happen. There are many sources of energy. Sun, fossil fuels, chemical reactions, wind, water, atoms, and molten rock are some of our energy sources. Energy cannot be created or destroyed, but any form of energy can be changed into another form of energy. In the following design briefs, you are going to explore how we use different forms of energy to provide the power we need.

Design and make a pendulum on a freestanding frame that will allow you to investigate motion. Determine what you must do to

- increase the height of the swing
- extend the period, or time, required for a full swing

Use tables to keep track of your trials. What source of energy is used by a pendulum? Can you think of a piece of playground equipment that is a pendulum?

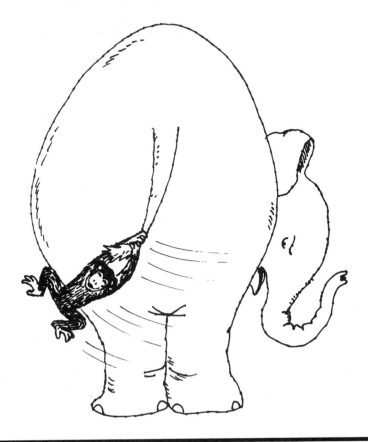

Teaching Others

The second-grade class is learning about transportation. Your class has been asked to demonstrate an air-powered gondola that travels on a string track.

Design and build a gondola that is powered by a balloon. Your gondola should be able to travel along 6 to 9 feet of string, carrying the heaviest load possible.

Things to consider

- What kind of material should be used?
- What is the best angle at which to attach your string in order to improve distance?
- How much air should you put in the balloon?
- What is the best angle at which to attach your string in order to increase the weight that can be carried?
- What form of energy is used to power the gondola?
- Can you explain to the second graders how the gondola works?

Let It Blow

The wind is a very inexpensive source of energy. By harnessing the wind's energy, you can produce electricity. In this design brief, your task is to design and construct a windmill which will power a motor and then light a bulb. You can use a fan as a source of wind to test your windmill.

Let It Shine

The energy released from the sun provides almost all of the power used on earth. It is the ultimate source of most other forms of energy. Collecting solar energy and realizing its potential is critical to our survival. In this design brief, your task is to do just that.

Design and construct a collector can for solar energy. Your collector must be able to maintain a temperature that is at least 10 degrees warmer than the temperature in your classroom. The real challenge will be choosing your insulator and thinking of ways to make the can collect the heat of the sun's rays. Think about how you are going to monitor the temperature. Continue testing and revising your design until your collector can maintain the required temperature.

Energy Design Brief 5
Fun and Games

Electricity is a form of energy that has improved the quality of our lives, perhaps more than any other. Imagine what the world was like before electricity was discovered. Make a list of all the things you would have to do differently without electricity. Even our forms of entertainment have been affected by electricity. What about education? How has electricity affected that? This design brief will allow you to improve some aspect of education by using electricity. Design and construct an educational game that can be used to teach or review a concept. Your game should be in the form of a hidden circuit board that lights up or buzzes when a player gets the correct answer.

Energy Design Brief 6
Pick Me Up

Whoops! One of the kindergarten students just dropped a whole box of paper clips into their sandbox. The child has come to your class for help. The teacher says children cannot play in the sandbox again until every single paper clip has been recovered. There are 100 of them! You think about it for a moment. A magnet would be great, but you do not have one. What you do have is wire, a nail, a battery, and some materials that you have gathered from the classroom. Use these materials to design and make an electromagnet that is capable of picking up as many paper clips as possible.

Things to consider.

- What material should you use for the core?
- How many times should you coil the wire around the core?
- What makes the electromagnet work?
- What makes the wires get hot when you are using it?

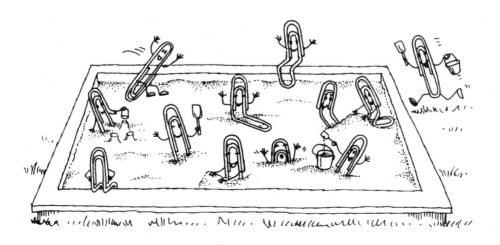

KidTech © Dale Seymour Publications®

Hover Power

Your school has organized vehicle races in the gym as part of your school fair. All vehicles entered must be made by students and they must move without wheels. Prizes will be given for the most attractive vehicle, the fastest vehicle to travel a distance of 3 yards, and the vehicle that can travel the farthest without stopping.

Design and make a vehicle that will hover across the floor. Your vehicle should be constructed with the intention of winning one of the prizes offered. Which prize will you go after? Maybe you can win more than one!

Tourist Attraction

You are just about to open your own water park near one of the largest waterfalls in the world. You have one thing left to do: make a sign to advertise your park. You really want your sign to use water to reflect the theme of the park. Design and construct a model of a freestanding waterwheel that can be used to turn a sign advertising the activities your park has to offer. Your waterwheel will be powered by the water from the falls, so it will be inexpensive too. Your sign should attract customers!

Gravity Power

Potential energy is all around us. Our society has learned to harness many forms of potential energy, such as the wind, water, fossil fuels, and the sun, and turn it into kinetic energy that will do work for us. In some clocks, the potential energy of a falling weight is used to supply kinetic energy to make the clock parts move. Your task in this design brief is to design and make a car that can be powered by a falling weight. Your car frame should be simple, be made of lightweight wood, and have a triangular tower from which you can attach your weight. Once the weight is released, your car should be able to travel a distance of at least 3 feet or go up a ramp 2 inches high and 2 feet long.

Now We're Cooking

You are trying to teach a four-year-old about how powerful the sun is without hurting him. You decide the best way to do this is by using toasted marshmallows! Design and build a solar cooker that will allow you to toast marshmallows for your little buddy. What safety features must you consider?

Message from Above

You have been grounded and are not allowed to use the phone or leave your fortieth-floor apartment. You have access to the balcony and have tried calling to your friends on the street, but they cannot hear you. You need to let them know you can't come to the movies with them. You decide to drop a message down to them.

Design and build a parachute that you can use to drop your message.

Things to consider

- What forces will affect your parachute once you drop it?
- Will wind make a difference?
- Will the weight of the load make a difference?
- Will the size and shape of the canopy make a difference?
- How will air pass through the canopy so that it will descend and not float away?
- Will the length of the string make a difference?
- How will you make sure that your parachute attracts the attention of someone down below?
- Will your parachute hurt anyone if it hits them? (You would not want that to happen!)

Transportation Design Brief 2
Ready! Aim! Fire!

A catapult is a weapon that was used in medieval times. A catapult is useful for moving things from one place to another. A catapult can help you move something across water, a ravine, a street, or piece of land you are unable to cross over yourself.

Design and construct a catapult that can be used to move a load of marshmallows a distance of at least 3 yards. Your catapult must contain a measuring device that will allow you to regulate the firing distance.

Fly with Us

The airline industry is very competitive. Imagine that you have just started your own airline. Prepare a one-minute TV commercial to convince people to fly with your airline.

Things to consider

- What will you call your airline?
- Will you have a slogan?
- Will you have a logo?
- What will your colors be? Can you use them in your ad?
- What promotions or bargains will you offer?
- What makes your airline better than others?

When you have prepared your commercial and practiced it, perform it for the class or videotape it. If you have the opportunity to view your video, decide if you would change it in any way.

Space Robot

Robots are often used to transport goods. They are capable of doing work such as carrying or lifting heavy loads. They are especially useful for performing tasks that are unsafe, too difficult, or too monotonous for humans.

Design and make a robot that can perform one task. Give your robot a name and write instructions for the person who will be operating the robot.

SNORE

Wanted: Space Project Workers

Have you ever thought about what you would like to be when you grow up? Have you ever wondered how the career you choose may change by the time you are old enough to do it, or what new careers might be available? When your grandparents were your age, not too many people thought about being a computer technician or an astronaut. Imagine a career that has something to do with space travel. Design a Help Wanted advertisement for that career. You may want to do your final copy on a computer.

Things to consider

- What will you call the position?
- What will the job involve?
- What education or training will the applicant need?
- What personal characteristics are necessary?
- Will the successful candidate have to relocate?
- What will be the salary and benefits?

I Missed My Train

Train stations are busy places. Thousands of people pass through every day. People are leaving on trains, arriving on trains, seeing people off, picking people up, or just working at the train station. Most people never stop to think about how complicated a train station really is. You would, though, if your job was to design a train station! That is your job today.

Design and construct a model train station for a city of one million people.

Things to consider

- You will need restrooms. What other services will you have to provide?
- Where will people park their cars; what traffic control is needed?
- What trains and equipment are needed?
- Will there be different ticket windows for different trains?
- How will baggage be handled?
- What security will you provide?

Blast Off!

A thousand years ago, people would have laughed at you if you dreamed about flying. In today's world, flying is as common as walking or driving. It is still a big event, however, when people fly in space. By the time you are an adult, flights to space stations will probably be taking place regularly. Some of you will probably fly in space, work on space stations, or build spaceships. Just as there are many models of cars today, there will be many models of spaceships.

Design and make a model of what you would like a spacecraft in the year 2020 to look like. You may use only materials you have gathered. When you have completed your model, make a commercial for your spaceship to encourage sales.

Puff Power

One of the challenges in designing a vehicle is deciding how the vehicle will be powered. Balloons can be used to provide jet propulsion for a lightweight vehicle. As the air escapes from an inflated balloon, it pushes the vehicle in the opposite direction. Design and construct a lightweight racing car that can be powered by a balloon.

Things to consider

- What materials will ensure a light weight car?
- How can you keep air from escaping until you are ready to run your car?
- How can you control the flow of air so that it comes out at a certain speed?

Move It and Dump It

Your class has been given the task of moving the sand from the sandbox in one classroom to a new sandbox that has just been built in the classroom across the hall. You decide to have some fun doing it and, at the same time, see what it would be like to run your own business. Design and build a pneumatically operated dump truck with a tailgate that will open to release the sand. Set up a trucking company that can be used to complete the job. Things to consider

- Time is money. How will you schedule the job?
- Remember, your reputation is based on how efficiently you work.
- What skills will your workers need?
- Do you have sufficient staff to do this job efficiently?

Public Transit

To design public transportation vehicles, you must consider things you don't need to think about if you're designing private vehicles. You must consider the cost of building the vehicle compared to how much income it can generate. You must consider the safety and comfort of many passengers. The inside layout of the vehicle also requires creative planning if you are going to maximize its potential.

Design and construct a bus that can transport twenty people comfortably. Your bus does not have to look like the buses you see everyday; you may be able to come up with a better design. Can you make a door that opens and closes?